The Globotics Upheaval

The Globotics Upheaval

Globalization, Robotics, and the Future of Work

RICHARD BALDWIN

Oxford University Press is a department of the University of Oxford. It furthers
the University's objective of excellence in research, scholarship, and education
by publishing worldwide. Oxford is a registered trade mark of Oxford University
Press in the UK and certain other countries.

Published in the United States of America by Oxford University Press
198 Madison Avenue, New York, NY 10016, United States of America.

Library of Congress Cataloging-in-Publication Data
Names: Baldwin, Richard E., author.
Title: The globotics upheaval : globalization, robotics, and the future of
work / Richard Baldwin.
Description: New York, NY : Oxford University Press, [2019] | Includes index.
Identifiers: LCCN 2018012182 (print) | LCCN 2018013452 (ebook) |
ISBN 9780190901776 (UPDF) | ISBN 9780190901783 (EPUB) |
ISBN 9780190901769 (hardback)
Subjects: LCSH: Employees—Effect of technological innovations on. |
Robotics—Economic aspects. | Automation—Economic aspects. |
Globalization—Economic aspects. | BISAC: TECHNOLOGY & ENGINEERING /
Robotics. | BUSINESS & ECONOMICS / Development / General. | BUSINESS & ECONOMICS /
Office Automation.
Classification: LCC HD6331 (ebook) | LCC HD6331.B254 2019 (print) |
DDC 331.25—dc23
LC record available at https://lccn.loc.gov/2018012182

9 8 7 6 5 4 3 2 1

Printed by Sheridan Books, Inc., United States of America

CONTENTS

The Globotics Upheaval

1

Introduction

Hang gliding is the ultimate thrill sport, but it's not as dangerous as you might think—thanks to the US Hang Gliding and Paragliding Association (motto: "Pilot safety is no accident"). To set up an online accident reporting website, the Colorado-based association signed a contract with California company Hathersage Technologies. The trouble was that Hathersage didn't have employees with the necessary skills.

Francis Potter, Hathersage's president, wasn't worried. He planned to recruit all the talent he needed within days, and pay them far less than the going wage. This was not foolish optimism. Potter had a secret up his sleeve. Using a web platform called Upwork, which is something like eBay for freelancing, he hired engineers from Lahore, Pakistan, to help him do the job. Potter is a big fan of foreign freelancers.

"There are really talented people who are just looking for the right opportunity to help on interesting projects. Upwork allows ordinary businesses to tap into latent capability and energy all over the world, whether in a basement in Siberia, a family house in Cambodia, or a small office in Pakistan," he wrote.[1]

If you look this straight in the eyes, you'll see it for what it is. It is US workers facing direct, international wage competition. It is highly skilled,

1. Francis Potter, "How the Hathersage Group Built a Global Development Team," *Upwork* (blog), September 21, 2016, https://www.upwork.com/blog/2016/09/hathersage-group-global-development-team/.

low-cost foreign workers working (virtually) in US offices. Using foreign-based freelancers may not be quite as good as using on-the-spot workers, but—as Potter can attest—it is a whole lot cheaper.

Think of this as telecommuting gone global. Think of it as telemigration.

TELEMIGRANTS—NEW PHASE OF GLOBALIZATION

These "telemigrants" are opening a new phase of globalization. In the coming years, they will bring the gains and pains of international competition and opportunities to hundreds of millions of Americans and Europeans who make their living in professional, white-collar, and service jobs. These people are not ready for it.

Until recently, most service and professional jobs were sheltered from globalization by the need for face-to-face contact—and the enormous difficulty and cost of getting foreign service suppliers in the same room with domestic service buyers. Globalization was an issue for people who made things; they had to compete with goods shipped in containers from China. But the reality was that few services fit into containers, so few white-collar workers faced foreign competition. Digital technology is rapidly changing that reality.

Way back in the old days—which means 2015 on the digitech calendar—the language barrier and telecom limits restricted telemigration to a few sectors and source countries. Foreign freelancers had to speak "good-enough English," and they were limited to modular tasks. Telemigrants were common in web development, and a few back-office jobs, but little else. Things are different now in two ways.

Machine Translation and the Talent Tsunami

First, machine translation unleashed a talent tsunami. Since machine translation went mainstream in 2017, anyone with a laptop, internet connection, and skills can potentially telecommute to US and European offices.

This is amplified by the rapid spread of excellent internet connections. This means that people living in countries where ten dollars an hour is a decent middle-class income will soon be your workmates or potential replacements.

Chinese universities alone graduate eight million students a year, and many of them are underemployed and underpaid in China. Now that they can all speak "good-enough English" via Google Translate and similar software, special people in rich nations will suddenly find themselves less special.

Think about that. Then think about it again.

This international talent tidal wave is coming straight for the good, stable jobs that have been the foundation of middle-class prosperity in the US and Europe, and other high-wage economies. Of course, the internet works both ways, so the most competitive rich-nation professionals will find more opportunities, but for the least competitive, it is just more wage competition.

Second, telecom breakthroughs—like telepresence and augmented reality—are making remote workers seem less remote. Widespread shifts in work practices (toward flexible teams) and adoption of innovative collaborative software platforms (like Slack, Asana, and Microsoft 365), are helping to turn telemigration into tele-mass-migration. And there is more.

This new competition from "remote intelligence" (RI) is being piled on to service-sector workers at the same time as they are facing new competition from artificial intelligence (AI). In short, RI and AI are coming for the same jobs, at the same time, and driven by the same digital technologies.

WHITE-COLLAR ROBOTS—NEW PHASE OF AUTOMATION

Amelia works at the online and phone-in help desks at the Swedish bank, SEB. Blond and blue-eyed, as you might expect, she has a confident bearing softened by a slightly self-conscious smile. Amazingly, Amelia also works in London for the Borough of Enfield, and in Zurich for UBS. Oh, and did I mention that Amelia can learn a three-hundred-page manual in thirty seconds, can speak twenty languages, and can handle thousands of calls simultaneously?

Amelia is a "white-collar robot." Amelia's maker, Chetan Dube, left his professorship at New York University convinced that using telemigrants from India would be nowhere near as efficient as replacing US and European workers with cloned human intelligence. With Amelia, he thinks he is close.

If you look this straight in the eyes, you'll see it for what it really is. It is zero-wage competition from thinking computers. Amelia and her kind are not enhancers of labor productivity—like faster laptops, or better database systems. They are designed to replace workers; that's the business model. Amelia and her kind are not quite as good as real workers, but they are a whole lot cheaper, as SEB can attest.

These thinking computers are opening a new phase of automation. They are bringing the pluses and minuses of automation to a whole new class of workers—those who work in offices rather than farms and factories. These people are unprepared.

Until recently, most white-collar, service-sector, and professional jobs were shielded from automation by humans' cogitative monopoly. Computers couldn't think, so jobs that required any type of thinking—be it teaching nuclear physics, arranging flowers, or anything in between—required a human. Automation was a threat to people who did things with their hands, not their heads. Digital technology changed this.

A form of AI called "machine learning" has given computers skills that they never had before—things like reading, writing, speaking, and recognizing subtle patterns. As it turns out, some of these new skills are useful in offices and this makes white-collar robots like Amelia into fierce competitors for some office jobs.

The combination of this new form of globalization and this new form of robotics—call it "globotics"—is really something new.

The most obvious difference is that it is affecting people working in the service sector instead of the manufacturing and agricultural sectors. This matters hugely since most people have service-sector jobs today. The other differences are less obvious but no less important.

WHY THIS TIME IS DIFFERENT

Automation and globalization are century-old stories. Globotics is different for two big reasons. It is coming inhumanly fast, and it will seem unbelievably unfair.

Globotics is advancing at an explosive pace since our capacities to process, transmit, and store data are growing by explosive increments. But what does "explosive" mean? Scientists define an explosion as the injection of energy into a system at a pace that overwhelms the system's ability to adjust. This produces a local increase in pressure, and—if the system is unconfined or the confinement can be broken—shock waves develop and spread outward. These can travel "considerable distances before they are dissipated," as one scientific definition dryly described the devastating blast wave.[2]

Globotics is injecting pressure into our socio-politico-economic system (via job displacement) faster than our system can absorb it (via job replacement). This may break the societal confinements that restrain hostility and violent reactions. The result could be blast waves that travel considerable distances before they dissipate.

Deep down, the explosive potential comes from the mismatch between the speed at which disruptive energy is injected into the system by job displacement and the system's ability to absorb it with job creation. The displacement is driven at the eruptive pace of digital technology; the replacement is driven by human ingenuity which moves at the leisurely pace it always has.

The radical mismatch between the speed of job displacement and the speed of job replacement is the real problem. The direction of travel is not. Service-sector automation is inevitable and welcome in the long run.

2. Elain S. Oran and Forman A. Williams, "The Physics, Chemistry, and Dynamics of Explosions," *Phil. Trans. R. Soc. A*. 370, no. 1960 (2012): 534–543, http://rsta.royalsocietypublishing.org/content/roypta/370/1960/534.full.pdf.

But why is this technological impulse so much faster than those that transformed the economy from agrarian to industrial, and from industrial to services? The answer, strange as it may seem, lies in physics.

A Very Different Physics

Past globalization and automation were mostly about goods—making them and shipping them. They were thus ultimately restrained by the laws of physics that apply to goods (matter). Globalization and automation of the service sector are all about information (electrons and photons)—processing them and transmitting them. Globotics is thus ultimately linked to the laws of physics that apply to electrons and photons, not matter. This alters possibilities.

It would be physically impossible to double world trade flows in eighteen months. The infrastructure could not handle it, and building infrastructure takes years, not months. World information flows, by contrast, have doubled every couple of years for decades. They will continue to do so for years to come.

The timescale disparity is due to differences in the relevant physics. Electrons can violate many of the laws of physics that slow down globalization and automation in industry and agriculture. This is one reason that today's technological impulse is profoundly different than the technological impulses that triggered previous waves of automation and globalization. This is why historical experience must be treated with great care when applying lessons to today's globalization and robotization. And it is exactly why the disordering of service-sector jobs will come faster than most believe.

But speed is only the first big problem. The second is the fact that America's and Europe's middle classes will come to view both types of globots—telemigrants and white-collar robots—as unfair competitors.

Outrageously Unfair

Nothing makes people angrier and more prone to violent reactions than unfair competition. Sociologists tell us that people can keep a "cap on their crazy" when they are embedded in a social matrix of rules and restraints. When everyone plays by the rules, we can all play the game. But when some of the rules are broken, the cork can come out of the crazy, and more rules get broken.

Consider this in the light of the globalization part of globots.

Unlike the old globalization, where foreign competition showed up in the form of foreign goods, this wave of globalization will show up in the form of telemigrants working in our offices. We will see their faces and know their stories. This will be humanizing but won't change the basic fact that they will undermine our pay and perks.

These new competitors will accept lower pay at least in part because they won't pay the same taxes or face the same costs of housing, medical care, schooling, or transportation. They won't be subject to the same labor laws or workplace regulations. They won't ask for severance pay, paid holidays, pension contributions, or maternity and paternity leave. They won't pay taxes that support social security, social medical insurance, or any other social policies.

The ability of Americans and Europeans to ask for these benefits will inevitably be curtailed by the fact that telemigrants won't ask for them. The robot part of globots will be unfair in similar ways.

White-collar robots are paid zero wages and they are incapable of accepting perks. You cannot force a "cogitating computer" to take holidays, lunch breaks, or sick days. They aren't subject to workplace regulations, and they'll never join a union. They can work 24/7 if need be and be cloned without limits. The industry calls them "digital workers," but in fact they are nothing more than computer software.

To put it directly, competition from software robots and telemigrants will seem monstrously unfair. And this is why it will be easy for populists to characterize globots as unscrupulous efforts by large corporations to undermine the bargaining power of American and European service-sector workers.

Due to the logic of workplace competition, the very existence of telemigrants and cognitive computers will undermine workplace protections, benefits, and wages. Perhaps they already are.

THE GLOBOTICS UPHEAVAL

In today's job-centric capitalism, prosperity is based on good, secure jobs—and the stable communities that are built on them. Many of these jobs are in the sectors that globots will disrupt. And we are talking about a lot of jobs.

Estimates of the job displacement range from big—say one in every ten jobs, which means millions of jobs—to enormous—say six out of ten jobs, which means hundreds of millions. When millions of jobs are displaced and communities are disrupted, we won't see a stay-calm-and-carry-on attitude.

Backlash Bedfellows

The Trump and Brexit voters who drove the 2016 backlash know all about the job-displacing impact of automation and globalization. For decades, they, their families, and their communities have been competing with robots at home, and China abroad. They are still under siege financially. Their futures look no brighter. The economic calamity continues—especially in the US. For these voters, the policies adopted in the US and UK since 2016 are the economic equivalent of treating brain cancer with aspirin. Many populist voters also feel their communities are still under fire culturally. All that the Trumps and Brexiteers have provided is more "bread and circuses" to sooth the soul and primp the pride.

These populist voters will still be yearning for big changes in 2020. And they will, I believe, soon have a lot of company.

The urban, educated people who voted against populism will have a whole new attitude when globalization and automation get up close and personal. Professional, white-collar, and service-sector workers will seek to slow or reverse the trend. They will clamor for shelter from the globots. Perhaps the movement will come to be called "shelterism"—not antiprogress, just a little shelter from the storm.

In this scenario—which is just one of many—people who were on opposite sides of the "Trump fence" in 2016 will find themselves on the same side of a very different fence in 2020. One precedent is the way that the antiglobalization movement of the 1990s combined very different, and previously opposed, groups—environmentalists on one hand, and labor unionists on the other hand. We can't know what "fence" will define the globotics upheaval. Maybe it will be an antiglobotics fence, an antitechnology fence, or an anticorporate fence. Or maybe voters will just be angry in isolation so it becomes a free-for-all melee. The complexity of political dynamics makes these things impossible to predict, but we can already see hints of what is to come.

Many people in advanced economies already share a sense of outrage, urgency, and vulnerability. When white-collar workers start sharing the same pain, some sort of backlash is inevitable. All that is needed is a populist politician to capture their imagination. In fact, there already is a populist trying to unite blue-collar and white-collar anger: Andrew Yang.

Yang, who already entered the 2020 presidential race, argues that the US needs radically new policies to prevent mass unemployment and a violent backlash. "All you need is self-driving cars to destabilize society . . . That one innovation will be enough to create riots in the street. And we're about to do the same thing to retail workers, call center workers, fast-food workers, insurance companies, accounting firms."[3] Yang is—as *New York Times* writer Kevin Roose puts it—"a longer-than-long shot"

3. Kevin Roose, "His 2020 Campaign Message: The Robots Are Coming," *New York Times*, February 10, 2018, https://www.nytimes.com/2018/02/10/technology/his-2020-campaign-message-the-robots-are-coming.html.

presidential candidate, but his themes are likely to be taken up by more electable candidates. "If we don't change things dramatically," Yang says in his "Andrew Yang for President" video, children will grow up in a country with "fewer and fewer opportunities and a handful of companies and individuals reaping the gains from the new technologies while the rest of us struggle to find opportunities and lose our jobs."

This is something we should all worry about. We don't know what the pushback will look like, but as the *Game of Thrones* character, Ramsay Snow, said so aptly: "If you think this has a happy ending, you haven't been paying attention."

The Upheaval and Backlash

The last great upheaval—industrialization's rapid and unguided progress in the nineteenth century—created a world where job loss meant poverty and perhaps starvation for landless workers. While we did eventually learn how to make industrialization work for the majority, the process was spread over two world wars and the Great Depression. Individuals and countries across the world embraced fascism and communism as part of the backlash. People elected populists who promised authority, justice, and economic security—just as they do today.

Any new upheaval—the globotics upheaval, if you will—could spread very quickly since globots are really a worldwide challenge. To avoid such extremes, our governments need to ensure that globotics seem more like a decent development than a divisive disintegration. The new phases of globalization and robotics need to be seen by most people as fair, equitable, and inclusive. We need to prepare.

Preparing for the Upheaval—Protect Workers, Not Jobs

There is nothing wrong with globotics' direction of travel—it's the speed and the unfairness that pose the problems. Governments need to help workers adjust to the job displacement, foster job replacement, and—if the pace turns out to be too great—slow it all down.

The first step is to reinforce policies that make it easier for people to adjust. No new policies are needed, just more of the adjustment policies that have worked in Europe—things like retraining programs, income support, and relocation support.

The second step is to find a way to make the rapid job displacement politically acceptable to a majority of voters. Governments who want to avoid explosive backlashes must figure out how to maintain political support for the changes that are coming in any case. Politics is a fine art involving inspiration and leadership as well as concrete policies, but whatever they use, our political leaders will have to find ways of sharing the gains and pains, or at least offering a perception that everyone has a fighting chance of being a winner.

While tax-and-redistribute policies undoubtedly have to be part of this package, they cannot be the only thing, or even the main thing. People's lives are too tied up with their jobs to allow it. The challenge is ensuring labor flexibility doesn't mean economic insecurity for workers. What is needed are policies like those in Denmark. The government allows firms to hire and fire freely but then commits to doing whatever it takes to help the displaced workers find new jobs.

The good news is that once we make it past the upheaval, the world will be a much nicer place.

A MORE HUMAN, MORE LOCAL FUTURE

Automation and globalization displaced jobs in the nineteenth and twentieth centuries. Human creativity—being boundless—invented "needs" that we did not even know we needed. That's why many of us today work in jobs that would sound very strange to Charles Dickens in nineteenth-century London. Imagine what he'd think if you told him his great-great-grandchildren would be web developers, life coaches, and drone operators?

The jobs were created in service sectors since they were the sectors that were shielded from automation and globalization. The same will happen again today. Jobs will appear in sheltered sectors. But what sort of jobs will these be?

We cannot know what new jobs will be, but by studying the competitive advantage of AI and RI, we can say quite a bit about what sheltered jobs will look like in the future. By taking a close look at what RI does well, it is clear that the jobs that survive competition from telemigrants will be those that require face-to-face interactions. Psychologists have studied why in-person meetings are so different than email, phone, or Skype. The "secret sauce" for why real face time is so much more valuable is complex, and based on evolutionary forces that shaped our brains over millions of years.

While digitech is creating ever better substitutes for being there, it seems that for many years, "being there" will still matter for some types of workplace tasks. The jobs that survive and the new ones that arise will involve a lot of such tasks. The implication of this point is straightforward. These jobs will make our communities more local, and probably more urban.

By studying the things that AI-trained robots like Amelia can already do well, we can predict that the jobs that survive competition from AI and the new jobs that will be created are those that stress humanity's great advantages. Machines have not been very successful at acquiring social intelligence, emotional intelligence, creativity, innovativeness, or the ability to deal with unknown situations.

Experts estimate that it will take something like fifty years for AI to attain top-level human performance in social skills that are useful in the workplace, like social and emotional reasoning, coordination with many people, acting in emotionally appropriate ways, and social and emotional sensing. This suggests that the most human skills will be sheltered from AI competition for many years. The implication is as simple as it is profound. Humanity will be important in most of the jobs of the future.

All this, taken together, is why I am optimistic about the long run, why I believe the future economy will be more local and more human.

The sheltered sectors of the future will be those where people actually have to be together doing things for which humanity is an edge. This will mean that our work lives will be filled with far more caring, sharing, understanding, creating, empathizing, innovating, and managing people who are actually in the same room.

This is a logical inevitability—everything else will be done by globots.

While I believe this happy finale is where digital technology will take us ultimately, it is not the right place to start our reflections on the changes that are coming. The place to start is the past. The passcode to understanding the future is hidden in the lessons of history.

GLOBOTICS TRANSFORMATION AS A FOUR-STEP PROGRESSION

The massive changes that are coming will involve insanely complex interactions between technological, economic, political, and social forces. To put some order in this complexity, it is useful to group the changes into a four-step progression—transformation, upheaval, backlash, and resolution—all of which are launched by a technological breakthrough.

"Step" here is not meant in a sequential sense. The transformation, upheaval, and backlash can all develop at the same time, and the resolution need not put an end to it. That is how it happened in the past.

Two Historical Tech Impulses and Transformations

The Globotics Transformation will be the third great economic transformation to shape our societies over the past three centuries. The first— known as the Great Transformation— switched societies from agriculture to industrial and from rural to urban. This started in the early 1700s. The second, which started in the early 1970s, shifted the focus from industry to services. I call it the "Services Transformation" to contrast it with the industrial transformation that preceded it. Today's Globotics Transformation is focusing primarily on the service sector. It will shift workers to service and professional jobs that are "sheltered" from telemigrants and white-collar robots.

The three technological impulses that launched these are very different and thus had very different effects.

Oversimplifying to make the point, the Great Transformation was launched by the Steam Revolution and all the mechanization that followed. This technology took the horse out of horsepower; it created better tools for people who worked with their hands as Erik Brynjolfsson and Andrew McAfee point out in their seminal 2014 book, *The Second Machine Age*.[4] It was mostly about goods, and it shifted the masses from making farm goods to making manufactured goods. Office work grew more productive, but mostly due to the fruits of industrialization (office machinery, electricity, etc).

The Services Transformation was launched, in 1973, by the development of computers-on-a-chip and all the Information and Communication Technology (ICT) that followed. This technological impulse pushed the economy in a radically different direction, since it was radically different— Byrnjolsson and McAfee call it the Second Machine Age.

ICT created better substitutes for people whose jobs involved manual tasks and better tools for people whose jobs involved mental tasks. The result was a "skills twist." The technology created jobs for people who worked with their heads but destroyed jobs for those who worked with their hands. The resulting deindustrialization devastated communities and created enormous social and economic difficulties for blue-collar workers—especially in nations that failed to help their citizens make the transition (like the US and UK).

The Globotics Transformation has been launched by a third technological impulse—digital technology. The digitech impulse is radically different than steam power and ICT, but in a way that is subtler than the difference between steam and ICT.

4. Erik Brynjolfsson and Andrew McAfee, *The Second Machine Age: Work, Progress, and Prosperity in a Time of Brilliant Technologies* (New York: Norton & Company, 2014).

When computers and integrated circuits started getting useful in the 1970s, automation crossed a "continental divide" of sorts. There are many ways of characterizing this crossing—a shift from things to thoughts, from hands to heads, from manual to mental, from brawn to brains, and from tangible to intangible. But regardless of how we think of it, computers could do only a highly restricted type of thinking. In fact, they weren't thinking in any real sense; they were just following an explicit set of instructions called a computer program. They were strictly obedient to the computer code.

Digital technology has pushed computing across a second "continental divide." Think of it as the switch from conscious-thought to unconscious-thought. Computers used to only be able to think in analytic, conscious ways since we only knew how to write computer programs that followed this type of thinking. Computers could not do intuitive, unconscious thinking since we didn't understand how humans think intuitively (we still don't).

A breakthrough in what is called "machine learning" allowed computers to jump over this limitation. Since 2016 and 2017, computers are as good or better than humans in some instinctual, unconscious mental tasks— things like recognizing speech, translating languages, and identifying diseases from X-rays.

Machine learning is giving computers—and the robots they run—new skills that are valuable in offices. Now they can mimic human thinking in tasks involving perception, mobility, and pattern recognition. Loosely speaking, machine learning is allowing computers to make choices that came "straight from the gut," as the legendary ex-CEO of General Electric, Jack Welch might say.[5]

The upshot of this new type of thinking computer is that automation is now affecting office jobs, not just factory jobs as in the past. The same digitech is also making it easy for foreign-based workers to perform tasks in our offices. It is making it seem almost as if these foreigners are actually in the room and speaking the same language.

5. Jack Welch and John Byrne, *Jack: Straight from the Gut* (Warner Business Books, 2001).

Another key difference between today's transformation and the last two concerns the timing. Globalization during the Great Transformation started one century after automation started. Globalization during the Services Transformation started two *decades* after automation. In today's Globotics Transformation, globalization and automation are taking off at the same time, and they are both advancing at an explosive pace.

Globalization and automation did wonderful things for us in the past, but the progress was paired with pain. In the future, they'll do a bit of both. Leveraging the future progress and alleviating the future pain will not be easy. But reviewing past upheavals should serve to guide our thinking.

PART I

Historical Transformations, Upheavals, Backlashes, and Resolutions

We've Been Here Before: The Great Transformation

Catherine Spence and her infant starved to death in the London Docklands. The year was 1869. A building boom had brought the Spences to London in the 1850s, but the 1866 financial crash bankrupted the shipyards. Her husband lost his job. The jobless had to choose between destitution-level local charity and the horrors of the workhouse. Catherine Spence went for the charity. Her starvation took two and a half years.

The Spences were caught up in the "Great Transformation," as twentieth-century thinker Karl Polanyi called it. This two-century sequence of incremental changes converted Europe from a collection of rural, farm-based economies ruled by monarchs to urban, industrial-based economies ruled by various flavors of democracy.

The Great Transformation was massively constructive—it created the modern world we live in today. It was also massively disruptive. A keyhole glimpse into the pain side of this gain-pain package comes from the inquest into Spence's death.

"They had pledged all their clothes to buy food, and some time since part of the furniture had been seized by the brokers for rent," the inquest noted. "The house in which they lived was occupied by six families . . . The jury on going to view the bodies found that the bed on which the woman and child had died was composed of rags . . . The windows were broken, and an old iron tray had been fastened up against one and a board up against another."[1]

People like the Spences—and the societies in which they lived—were unprepared for the new economic realities brought on by the "disruptive duo" of automation and globalization. The main problem was that the changes were so massive and, given the times, so fast. This makes the era an excellent source of historical lessons for today's upheavals in which the voracious velocity of job displacement is also the central problem. Lessons from the Great Transformation period, however, need to be handled with care. The changes back then involved a far more radical uprooting than anything America or Europe has seen recently, or is likely to see in the near future.

What Put the "Great" in the Great Transformation

For something like 120 centuries, civilization was supported by six inches of topsoil and regular rains. Prosperity for the masses was tied to having access to a bit of land; power for the elite was tied to taking a slice of that prosperity. As a result, the wealth of nations was founded on control of good agricultural land. There was trade and industry, but not much.

Moving anything anywhere was vastly expensive, very slow, and downright dangerous. It took Marco Polo, for example, three years to get from Italy to China; the return voyage took two years, and hundreds of his fellow travelers died on the way. Moving goods was less dangerous but no less difficult and expensive. Silk from China cost an emperor in Rome ten thousand times more than it cost in China.[2] Even ideas were difficult to move. Buddhism, for example, arose in India 2,500 years ago and took almost 1,000 years to get to China and Japan.

1. As described in John Ruskin, *Fors Clavigera: Letters to the Working Men and Laborers of Great Britain*, vol. IV (London: George Allen, 1874).

2. William Bernstein, *A Splendid Exchange: How Trade Shaped the World* (New York: Atlantic, 2008).

These constraints on moving goods, ideas, and people enforced a "dictatorship of distance" on all aspects of human life. With people tied to the land, almost everything had to be made within walking distance. The result was localism—the opposite of globalization. This spreading out of production across countless villages dominated the world's economic geography and dictated the realities of the pre-industrial world. On the upside, it gave us diversity. Centuries of localism, for example, is why there are over 5,000 brands of beer in Germany, and 350 grape varietals in Italy. It is why one language, Latin, evolved into different languages like Italian, Spanish, Portuguese, and French. The downsides were mostly economic.

The most important economic implication was stagnation. The tiny size of markets rendered innovation both difficult and not particularly valuable. And without innovation, there was no automation. Productivity stagnated. Living standards stagnated.

It wasn't just localism that kept the human condition in a wretched state. "Malthus's law" actively enforced misery. Even if a new swath of land, a new food crop, or a new plough were discovered, living standards rose, but only temporarily. In a generation or two, population pressure returned things to a state were most people were only one or two bad crops away from famine.

This was premodern growth. Economic expansion arose from employing more land and labor, not getting more out of each acre and hour. Income rose only until Malthus's diabolic feedback loop extinguished it.

Modern growth, which started in Britain in the late 1700s, is what repealed Malthus's diabolic law. Growth allowed each worker to produce a bit more every year, and this raised incomes year after year. By the twentieth century, most American and Europeans were miles away from starvation.

This is what put the capital "G" in the Great Transformation, but the transformation didn't come all at once. As mentioned, it is best thought of as a four-step progression: technology produces an economic transformation, the economic transformation produces an economic and social upheaval, the upheaval produces a backlash, and the backlash produces a resolution.

It's a great story.

TECHNOLOGICAL IMPULSE

Steam was hot stuff in the 1700s. The concentrated and controllable nature of the power, together with the fact that it was easily reproducible and eventually mobile, launched society onto a "happy helix"—a self-fueling, rising spiral where innovation drove industrialization; industrialization drove innovation; and both of them boosted incomes, which, in turn, fostered innovation and industrialization.

Steam power first got useful when the Newcomen engine started pumping water out of coal mines in Britain in 1712. It was not a sleek, high-tech marvel. It filled a three-story building, burned massive amounts of coal, and required constant tending, but it did one amazing thing. It took the horse out of horsepower. Newcomen's machine replaced hundreds of horses, and allowed miners to dig deeper and expand output while lowering costs. This was critical.

Coal was the lifeblood of the Great Transformation, so higher productivity in this sector was a key twirl in the happy helix's upward travel. The colossal shift of the population from country to city, and the economy from agriculture to industry required astronomical amounts of energy—amounts that would have been impossible to satisfy with firewood, water, and wind power.[3]

The next century and a half witnessed a "waltz" between steam power and mechanization. Steam engines got stronger, lighter and more fuel efficient as machine manufacturing got more precise. In turn, better steam engines made it easier and more worthwhile to develop better machinery. The process was cumulative. An especially notable milestone in this process came a half century after Newcomen took the horse out of horsepower. In 1769, James Watt's steam engine put the watt into wattage.

---

3. Just to meet British cooking and heating needs in 1860 with firewood would have require all of the nation's farmland turned into forests, according to Gregory Clark and David Jacks, "Coal and the Industrial Revolution, 1700–1869," *European Review of Economic History* (2007).

While this progress was revolutionary at the time—especially compared with the previous stagnation—it was slow by today's standards. It was nothing like the eruptive pace of the digital technology that is driving the Globotics Transformation. There was a century between Newcomen's engine and the first commercially viable steamships.

Revolutions are never just one thing. The steam impulse was matched by a very different but complementary impulse in the agricultural sector. It started with a land ownership shockwave called "enclosure."

British Agricultural Revolution

The British agricultural revolution started with the enclosure movement in the 1600s. This involved the fencing (enclosing) of land that used to be open. Enclosing land ended the access that many rural families had to lands formerly held in common (in the sense that any community member could graze animals on the land). The Boston Common—a big park in the middle of Boston—is one remaining example of a common that was established when Massachusetts was a colony of the British crown. Local farmers grazed cows there from 1630 until it became a public park in 1830.

When a common was enclosed, its use often switched to the main "cash cow" of the day—which turned out to be sheep, or more precisely the wool they produced. This drove people out of the agricultural sector since raising and sheering sheep commercially required far fewer workers than raising food for families. But it wasn't just switches in ownership that put the revolution in the agricultural revolution.

Enclosure firmed up property rights and thus encouraged adoption of more efficient farming techniques. One of the agricultural revolution's red-letter innovations was a switch to the four-crop rotation system that heightened the productivity of land. Improved farm machinery also accelerated productivity. The classic examples include automatic threshing machines for grain; seed drills for planting; and improvements in farming tools, like the switch from wooden to iron ploughs.

The upgraded tools and techniques made food cheaper and more abundant—an outcome that helped with a third impulse—a population explosion. The number of Brits doubled between 1750 and 1850.

The full list of things that were critical to getting the Great Transformation going is long and complex, but clarification is served by simplification. That's why it is insightful to focus on changes in British agricultural, population, and steam—especially steam.

TECHNOLOGY PRODUCES TRANSFORMATION

At first, steam technology mostly fostered mechanization and industrialization, or what we would call automation today. The trend started with the biggest industries of the time—textiles, coal, and iron—but it spread to other sectors over the decades.

Soon enough, the self-fueling spiral created a new linchpin industry— machine tools. Between 1770 and 1840, the British machine tool industry made great strides. This was a critical step since it lowered the cost of making the machines that helped automate production in general. The machine tool industry back then—like machine learning today—was a technology that accelerated technology's advance.

Before machine tools, industry really entailed what we would call handicraft. Rifles, for example, were constructed one at a time by highly skilled craftsmen using hand tools. Each rifle was unique (and thus expensive). Using machine tools, the American Eli Whitney standardized parts to such an extent that, from 1801, parts were interchangeable across his rifles. Production got faster and cheaper—partly because lower-wage, less-skilled workers could handle the work (an early example of the deskilling impact of technology).

This was a turning point in automation. Instead of highly skilled craftsmen making machinery out of wood and by hand, machine tools produced metal parts for machines that could be churned out with higher accuracy and lower costs. This sort of innovation cut both ways when it came to jobs.

Automation and Jobs—the Push and Pull Effects

Mechanization meant that the same pile of work could be done with fewer workers, but the cost savings also meant lower prices and thus more sales, and thus a higher pile of work. There was, in a sense, a race between the height of the pile and efficiency of workers. Call it the productivity-production foot race.

When the foot race was won by the piling-raising side—technology acted as a "pull factor"—it pulled workers into the sector. Where the efficiency side won, technology was a "push factor" pushing workers out of the sector. For example, enclosure, mechanization, and new farming techniques were massive push factors in the agriculture sector. The changes produced painful disruptions to livelihoods, families, and whole villages, but they released workers for jobs in industry and services.

There are important lessons in the way it happened. Technology eliminated many jobs but few occupations. The technology didn't eliminate the *occupation* of farming, for instance, it just meant that each farmer could feed more mouths, so fewer farmers were needed.

The mechanization of industry, by contrast, was a pull factor. While output per worker rose steeply, industial output rose even faster, so the number of workers in industry climbed.

A separate set of pull/push factors arose from the demand side. The most obvious dynamic was the way the booming population created more demand that created more jobs. A slightly subtler demand factor stems from the fact that people tend to change their purchase patterns as they get richer. At the income levels common at the time, people could afford very few goods. Some children went without shoes, and many adults wore second-hand clothes. As income rose above subsistence levels, people spent more on new goods, and the extra demand created extra manufacturing jobs.

Productivity itself was a demand factor for the very direct reason that if someone makes a thing, someone owns the thing. The thing thus becomes part of their income. Although the goods supplied and demanded could slip out of alignment temporarily, the general trend was for more output per worker to lead to more income per worker and more purchases per worker. Technically, this is called Say's law, which roughly corresponds to the notion that supply creates its own demand. Or, in the more rotund nineteenth-century phraseology of Jean-Baptiste Say: "As each of us can only purchase the productions of others with his own productions—as the value we can buy is equal to the value we can produce, the more men can produce, the more they will purchase."[4]

Globalization exaggerated both the push and pull factors in sectors that were open to trade. But the trade half of the tech-trade team lagged far behind. Steam power fired the starting gun on globalization a full century after Newcomen's steam engine unleased automation. The reason, quite simply, was that it took decades of refinements to make steam engines that were compact enough to put on wheels and ships.

Modern Globalization Starts

Railroads dramatically reduced the cost of moving goods. For the first time in history, the interiors of the world's great land masses were linked to the global economy. Steamships had an equally radical impact on seaborne transportation. The year 1819 saw the first steamship cross the Atlantic. The peace that came with the end of the Napoleonic Wars also gave globalization a mighty shove.

While traces of trade can be found back to the Stone Age, the early 1800s was the first time in history that the volume of trade really started moving the dial at the economy-wide level. For example, the whole of the 1600s saw only about three thousand European ships sailing to Asia and back, and the number wasn't much more than double

---

4. Jean-Baptiste Say, *A Treatise on Political Economy*, Grigg and Elliott, 1834; this translation from Guy Routh, *The Origin of Economic Ideas: Edition 2*, Springer, September 1, 1989.

that the whole of the 1700s. Each ship carried about a thousand tons of cargo.[5]

Oxford economist Kevin O'Rourke and Harvard economist Jeff Williamson date the beginning of modern globalization to 1820. This is when the price of, say, wheat inside Britain started to be set by international supply and demand conditions.[6] Before this date, food prices within a nation moved mostly according to changes in domestic supply and demand conditions—say, a crop failure or bumper crop. Once the volume of international trade was large enough, a crop failure would lead to lots of imports flowing into the country rather than the prices rising. This was an enormous change in the course of human events. For the first time, the ability to buy and sell goods internationally started having revolutionary effects on domestic economies.

None of this was sudden. Railroads recast land transportation, but the rail networks developed over decades. Steamships revolutionized ocean travel, but fueling problems prevented sole reliance on steam power for decades. For example, the first steamship that crossed the Atlantic combined wind and steam power due to fueling problems. The big switch came only after coaling stations had been set up all around the world.

The ability to sell to the whole world had massive effects on jobs. In Britain, where modern globalization first saw the light of day, it was a push factor for agriculture since food imported from the US and elsewhere was cheaper. Food imports boomed from the mid-1800s. But globalization is always a push-pull pair.

Jobs tend to move out of the sectors competing with imports, but move into sectors that export. In the case of the United Kingdom, booming imports of food were matched by equally booming exports of textiles and other manufactured goods.

---

5. See Angus Maddison, *Contours of the World Economy 1–2030 AD: Essays in Macro-Economic History* (Oxford: Oxford University Press, 2007) .

6. Kevin H. O'Rourke and Jeffrey G. Williamson, "When Did Globalization Begin?," *European Review of Economic History* 6 (2002): 2350.

The principle guiding this impact is David Ricardo's famous principle of comparative advantage, which, roughly put, says: "Do what you do best; import the rest." In nineteenth-century Britain, the "best" meant manufacturing. British competitiveness in manufacturing had a huge head start by the 1800s and its edge over other nations was still growing, so globalization allowed Britain to become the workshop of the world. The booming exports of manufactured goods kept the pile of work growing faster than efficiency of workers, and this pulled workers into industry.

The most dramatic impact of globalization, however, was the way it accelerated economic growth.

Modern Growth Starts

Modern growth—the sort of steady progress we arc used to today but was unheard of before the Industrial Revolution—depends upon innovation because more income requires more outcome. Achieving higher incomes every year requires that a nation's workforce produce more every year. That, in turn, requires that the workers have more or better "tools" every year. Here, "tools" mean capital broadly defined, namely human capital (which means skills, education, training, etc.), physical capital (which means machines, buildings, tools, etc.) or knowledge capital (which means technology, knowledge about production techniques, etc.). Of these three, knowledge is the key.

Knowledge capital is very different because innovation boosts the benefits of having more of the other forms of capital. Without innovation (or imitation of some other nation's innovations), investments in education and physical capital reach their limits and output per worker ceases to rise. Or, as economists phrase it, human and physical capital face diminishing returns, while knowledge capital does not. That is an empirical fact.

The reason is unclear, but one guess is that it reflects the fact that human ignorance is infinite despite millenniums of knowledge creation. Infinity is, after all, a concept not a number. Think of it as the biggest number you know plus one. And this means, infinite ignorance, even after you add a lot of knowledge, is still infinite.

Economically, the key is that innovation creates better processes for making old goods as well as brand new goods. This keeps economic growth rolling along. The century-long sequence of innovations in Victorian England are an excellent example. As innovations piled up, capital got more useful and thus continued to accumulate, as did human capital. Globalization entered the equation via its impact on innovation.

In the early 1800s, globalization boosted innovation in ways both simple and subtle. Exports lifted the constraint imposed by the size of the domestic market and this boosted the demand for innnovation. Selling to the world market also encouraged industries to concentrate geographically and this boosted the other side of the equation. With lots of people in the same place thinking about the same problems, the supply of innovative ideas rose. In short, innovation got easier just as selling to the world market made it more profitable. This is how the dynamic duo—automation and globalization—ignited the "bonfire" of modern growth. The bonfire is still burning.

Growth saw the ignition of a second-stage booster in the latter part of the 1800s. The acceleration was so marked that it has been given a name: the Second Industrial Revolution.

Technology Produces Technology—the Second Industrial Revolution

The happy helix, which had been spinning upward since the early 1700s, reached a new plateau in the second half of the 1800s. As machinery got more sophisticated, power got cheaper, and science was increasingly applied to industrial matters, a whole new group of industries sprung up. This created masses of new jobs for workers making things that had never existed—except in the science fiction novels of Jules Verne.

Robert Gordon, a professor of economics at Northwestern University, argues that the Second Industrial Revolution—what he calls the "special century" (1870–1970)—dropped a cluster bomb of innovations on the advanced economies. The economic "bomblets" exploded over a wide area, with each explosion producing a chain reaction of innovation, rising productivity, and income growth.[7]

This was an example of the happy helix of innovation and industrialization creating masses of new jobs in brand new sectors. Back then, as today, much of the job creation involved making things that were unthinkable only a few decades earlier. The new jobs were in making things related to railroads, telecommunications, electric lighting, internal combustion engines, and all types of electro-mechanical and electronic machinery including road vehicles, aircraft, radios and televisions, and industrial chemicals ranging from chemical fertilizers and herbicides to hair dyes and plastics.

These new industries were a long journey from cotton textiles. The developments, which were driven by automation and globalization, lighted the bonfire of sustained economic growth. Growth did wonderful things, but growth meant change, and change meant pain. The resulting gain-pain package led to the second aspect of the four-step progression, namely upheaval.

TRANSFORMATION PRODUCES UPHEAVAL

Oliver Twist—Charles Dickens's most memorable fictional character—could be a "poster child" of the upheaval. Born in a workhouse, Oliver is sold into apprenticeship at the age of nine after a thorough thrashing prompted by his famous, hunger-inspired, "Please, sir, I want some more."

Reality was almost as harsh for Charles Dickens himself. The second of eight children born into a middle-class family, Dickens was forced, at age twelve, to work in a factory when his father was thrown in debtors' prison. Things improved after the debt was paid and Charles returned to school,

---

7. The military analogies are mine. These inventions are sometime called the "Second Industrial Revolution," the first being mostly about textiles, steam and coal, and iron and steel.

but not for long. At fifteen, Dickens again had to take a job to help support his family.

Change brought pain—as it always does—and the faster the change, the greater the pain. The main avenues of change were fourfold: a shifting of workers out of agriculture and into industry, a shift of the population from farms to cities, a rise in inequality, and a shifting of the anchor of value creation and capture from land to capital.

Each change created its own gain-pain pairing and convulsed centuries-old social, economic, and political relationships. The traditional relationships were by no means idyllic, but they were what people were used to.

Urbanization: Linking Income Insecurity and Food Insecurity

When people moved from farms to cities, income security and food security got much more strongly linked than they had been in rural communities. Cities offered more opportunities than the countryside but this came at a cost. Industrial workers in cities had to buy all their food, so job loss was a life-threatening event. Even in the good times, wages for unskilled workers were low compared to the cost of living. Housing conditions were overcrowded and unsanitary; diets were poor; and accidents, sickness, or old age often led to deprivation, or even starvation.

Part of the fuel that stoked social strife in the Great Transformation came from the treatment of people who fell on hard times. Then, as now, many among the elite were quick to blame the misfortunate for their misfortune. British government policy at the time made things worse for the woeful, but it wasn't always that way in Britain.

Britain dodged the French Revolutionary "bullet," and not by accident. Geography was part of the explanation but also important was the "enlightened self-interest" of the landed elite, and earlier concessions made by the British monarchy to Parliament. Since the 1500s, a series of Poor Laws charged each local community (parish) with supporting its local poor. Systems varied regionally, but generally the support took the form of jobs, apprenticeships, or cash—all financed by taxes on the local well-off citizens, and overseen by local officials.

The "light" in enlightened self-interest dimmed considerably as the Great Transformation progressed and the booming population raised the cost of caring for the poor. Importantly, this extra burden fell especially hard on the urban elite since the poor were moving out of their country parishes and into the cities. The solution decided upon by the "good and the great" was a reform that would not look out of place in Trump's America. They made the Poor Laws poorer.

Contemporary critics of the traditional Poor Laws argued that the safety net encouraged people to have too many children, and generally seduced workers into laziness and dependency. They also encouraged employers to pay too little since workers could get public handouts. All this was to be fixed by the 1834 Poor Law Amendment. The 1834 act made it illegal to give support to people outside of workhouses, and then required the conditions in the workhouses to be horrible as a matter of moral principle. And it worked. Workhouses were widely feared—a terrible fate to be chosen only by the most desperate.

Victorian social thinkers like Reverend Thomas Malthus viewed poverty as a natural condition that particular workers fell into due to their personal moral failings. To avoid encouraging immorality and sloth, workhouse conditions were designed to be worse than those of the poorest free laborer outside of the workhouse. As Catherine Spence's example illustrates, such conditions shifted between fair-to-middling in good years to dire deprivation, or simple starvation, in downturn years.

Help receivers were stigmatized with special clothes and humiliated with strict rules; husbands and wives were separated to prevent families from growing. Work was mandatory and rations were meagre.

Income Inequality—The Ups and Downs

Almost as disturbing as the misery itself was the fact that prosperity was spreading as fast as the poverty. The affluent and the afflicted lived close together in Victorian London. The slums were built up in the same years as London's greatest attractions. Big Ben, the Victoria and Albert Museum, Marble Arch, and Trafalgar Square were all constructed in the decades bracketing Catherine Spence's starvation.

This contrast between the wealthy and the woeful made many view the massive social changes as outrageously unfair. Many thought the rich were getting richer because the poor were getting poorer. But what are the facts?

The real world that the fictional Oliver lived in was very unequal and inequality was growing. According to economic historians Peter Lindert and Tony Atkinson, inequality rose in the first part of the Great Transformation—say, up to the beginning of the Second Industrial Revolution.[8] After that, it declined right up to the end of the Great Transformation in 1970. The happy helix, in other words, was especially happy for the richest Britishers in its first century and especially happy for the middle class in its second century.

As Figure 2.1 shows, the share of income that went to the richest 5 percent in England and Wales rose gently from about 35 percent to about 40 percent during the first part of the Great Transformation—the so-called First Industrial Revolution, say 1759 to 1867.

The trend reversed in the late 1800s when the Second Industrial Revolution kicked in. Inequality fell quite dramatically in the UK as industrial growth got its second wind from the cluster of new industries. The income share of the top 5 percent dropped from 40 percent down to under 20 percent by the 1970s. Since then it's been rising, but that's a story for the next chapter.

It is not easy to say exactly what causes these waves of inequality. It is the subject of much debate, as Thomas Piketty's bestselling *Capitalism in the 21st Century* points out. By its very nature, inequality involves almost

8. See Max Roser and Esteban Ortiz-Ospina, "Income Inequality", published online at OurWorldInData.org, based on data from Peter H. Lindert, "When Did Inequality Rise in Britain and America?," *Journal of Income Distribution* 9 (2000): 11–25, and Anthony B. Atkinson, "The Distribution of Top Incomes in the United Kingdom 1908–2000," in *Top Incomes over the Twentieth Century: A Contrast between Continental European and English-Speaking Countries*, ed. Anthony B. Atkinson and Thomas Piketty (Oxford: Oxford University Press, 2007), ch. 4.

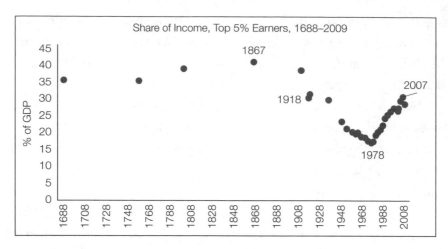

Figure 2.1 Income Inequality in the Great Transformation, 1688–2009.
SOURCE: Author's elaboration of data provided privately by Max Roser (*Our World in Data*). His sources are Peter Lindert "Three Centuries of Inequality in Britain and America," in *Handbook of Income Distribution*, ed. A. Atkinson and F. Bourguignon (Amsterdam: Elsevier, 2000); A. Atkinson, "The Distribution of Top Incomes in the United Kingdom 1908–2000," in *Top Incomes over the Twentieth Century. A Contrast Between Continental European and English-Speaking Countries*, ed. A. Atkinson and T. Piketty (Oxford: Oxford University Press, 2007); and B. Milanovic, P. Lindert, and J. Williamson, "Ancient Inequality," *The Economic Journal* 121, no. 551 (2008): 255–272, March 2011.

every aspect of the economic system—ranging from education, technology, and globalization to urbanization, voting rights, and imperialism. Most of these are interrelated.

A fair assertion, however, is that the initial upswing had to do with the rise of capitalism. Previously, landownership was the main way to get rich. The industrial revolution opened another important route—namely, capital ownership. This entailed both physical capital—like factories, ports, and ships—and financial capital—like ownership of stocks, bonds, and banks. All capital ownership is and always has been concentrated in the hands of the top 5 percent. Quite simply, only the rich could afford to save, so only the rich could build up their wealth, and their wealth helped them save and invest more, thus boosting their wealth. For the common people, incomes were spent fully on current consumption.

The other part of the equation is that wages grew more slowly than labor productivity. This can be understood as an issue of supply and demand. Rising labor productivity boosted the demand for labor, but the booming population growth and rural–urban migration meant that the supply rose even faster. Workers' ultimate alternative was to stay on low-income, low-productivity jobs in agriculture. To get a continual inflow of workers from the countryside, the industrial and urban wage had to be higher than the wage available on the farm, but they did not have to rise continuously.

The drop in inequality in the second phase reflects the fact that labor finally started getting scarce at the same time as the innovations started making labor especially productive. It is also surely important that this second phase corresponded, after World War I, with a rise in workers' negotiating and voting power.

In Britain, the power of unions rose in an uneven manner from just before World War I to the 1970s. The range of people who could vote expanded slowly although the 1800s, all men over age twenty-one and all women over thirty got the right to vote in 1918 (the discrimination was ended in 1928). Before that, men had to own a certain amount of property to vote—a restriction that tended to favor the political power of those who were already favored economically.

The Great Transformation was about much more than people changing jobs. The whole fabric of value (income) creation changed—along with the ways of capturing and controlling value.

Evolving Value Creation and Capture—Land to Capital

Before the Great Transformation, valuable economic things were mostly created by labor working on land. Laborers were abundant, and the supply could be increased via population growth. Land, by contrast, was more of a fixed factor. To own a bit of land was to control the value creation, and thus the value capture. This is why landowners controlled the division of the value created.

To line their own pockets, landowners only had to give the workers a large enough slice of the value to keep them alive and in place. That's why they called it feudalism: it was all about land. Land was the nucleus of the value creation. ("Feudalism" derives from the Latin word for a fief—a portion of land.) But land started to lose its center-point status with the rise of industry.

As the economic center of gravity shifted from farms to factories, value creation and capture also shifted. Land mattered much less. Capital became king. Manufacturing became the heart of modern economies. This, in turn, meant that capital working with labor became more central to income generation, that is, value creation. With much of the value created by labor working with capital, the focal point of economic value creation shifted from land to capital.

To own a bit of capital was to control the value creation, and thus the value capture. That's why it was called capitalism. Labor was still abundant, and capital wasn't really fixed, but capital owners were the ones with the power to decide the division of the value created. Of course, competition among capital owners constrained this power, but when one man— Henry Ford, for example—employed 100,000 workers, the power tended to be with the one rather than the many (until the many organized, but that is getting ahead of the timeline).

The shift in value creation and extraction can be seen very clearly in Figure 2.2, which shows the share of British income going to labor, capital, and land; and how the shares evolved from 1770 to 1910.[9] For a century following the beginning of the Great Transformation, capital's share rose. Land's share fell during the same hundred years, but its share continued to degrade even as capital's share of the "value-creation pie" stabilized.

9. The data is from Robert C. Allen, "Engel's Pause: A Pessimists Guide to the British Industrial Revolution," Department of Economics Discussion Paper Series no. 315, University of Oxford, April 2007.

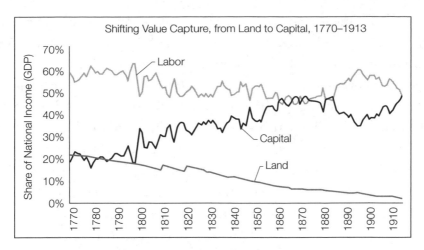

Figure 2.2 Capital and Land Shares of Value, 1770–1913.
SOURCE: Author's elaboration of data published in Robert C. Allen, "Class Structure and Inequality during the Industrial Revolution: Lessons from England's Social Tables, 1688–1867," *Economic History Review* 00, 0 (2018): 1–38.

UPHEAVAL PRODUCES BACKLASH

While glacial by today's standards, the changes proved too fast for nineteenth-century societies to absorb smoothly—especially as the rates of change accelerated toward the end of the century. The social pressure created by the speed was greatly amplified by a growing sense of injustice. The four massive changes—from farm to factory, from countryside to city, from land to capital, and rising inequality—ripped up the old rules and traditions that had long defined justice. Much of the backlash concerned conflicts over what the new rules should look like.

The novelty of the massive disruptions drove nineteenth-century thinkers to develop a whole new discipline aimed at understanding how social upheaval can lead to a backlash. It is called sociology. The founder of the new field was Émile Durkheim. Durkheim viewed people as inherently bent on chaotic selfishness. Social stability, he argued, was only possible because the socialization of individuals and their social integration held the underlying chaos in check. This view of social restraints could be called the "Durkheim Dike"—social order holds back individual chaos.

When economic and social upheaval broke enough of the constraints that had long held riot and mayhem in check, backlash was the result. And there was plenty of social disintegration going on. The shift from village life to overcrowded tenements in cities destroyed the social matrix of constraints stemming from family ties, religious rules, and the social hierarchy that people were used to. Durkheim's word for this state of socially unbound individualism is "anomie"—a lack of social and ethical standards. And other aspects of the Great Transformation violated key parts of the socialization rules that people had come to rely on.

One example is the Luddite Riots.

Small Backlashes in Britain

Revolt was in the air. The Napoleonic Wars had depressed the textile business, and poor harvests had generated high food prices and the occasional food riot. New, unsettling ideas from the 1789 French Revolution had drifted into northern England and were getting a hearing—things like human rights, government for and by the governed, and anti-monarchy sentiment.

Automation was thrown into this volatile mix in the form of the Cartwright power loom. It allowed an unskilled child to produce cloth three and half times faster than a skilled weaver using traditional technology. Weaver wages plummeted. Tens of thousands of weavers petitioned Parliament for a minimum wage—and were refused. Soldiers forcibly dispersed workers protesting for higher pay in Nottingham, and in reaction, the workers raided a nearby mill and hammered to pieces one of the new looms.

The year was 1811, and the moment became a movement. Loom-smashing spread and reactions turned violent. Workers, armed guards, soldiers, and mill owners died. But this backlash is widely misunderstood.

The Luddites were not primarily anti-technology. The skilled workers leading the upheaval were the nineteenth-century equivalent of today's unionized workers holding secure jobs with good pay and benefits. What they objected to was the way that automation allowed jobs that were traditionally reserved for qualified craftsmen to go to low-skill, low-wage workers—often young children. It just seemed outrageously unfair. It violated long-standing practices. It was something akin to the outrage provoked by the offshoring of American manufacturing jobs to Mexico. Repression was the instinctual reaction of the sitting government.

Parliament passed the Frame Breaking Act that allowed judges to impose the death sentence for loom-smashing. Over ten thousand troops were sent to quell the uprising. Dozens of protestors were hung and many more were transported to Australia. A similar movement arose against automation in farming (automated threshing machines). These so-called Swing Riots arose in southern England in the 1830s. They too were violently suppressed by the military and magistrates.

Globalization triggered a very different type of backlash.

The Napoleonic Wars hindered British imports in general and Continental grain imports in particular. This had boosted UK wheat prices and production—a delightful outcome for landowners. But when the war ended, grain imports surged and prices plunged. This triggered a backlash by aggrieved landowners. But they didn't have to hold rallies and break things. A simpler solution was at hand.

Large landowners held the reins of power in Parliament and engineered a protectionist backlash called the "Corn Laws." Passed in 1815, these laws raised prices of grain by keeping cheaper foreign grain out of Britain. This kept bread prices high for thirty years.

These British examples illustrate the general and very natural tendencies of great changes to generate great reactions. Similar things were happening on the Continent, but with a lag.

Failed Backlash on the Continent—1848

Continental Europe was not a business-friendly place in the years between the French Revolution (1789) and the end of the Napoleonic Wars (1815).

It was in a state of near continuous turmoil. When peace finally came, the old monarchies were restored by a set of deals known as the Congress of Vienna. This restored stability, and the stability bore economic fruit—it fostered the advance of automation and globalization. The stability, industrialization, and growth were welcome, but not enough. The Congress of Vienna and resulting growth did not redress the deep causes of the discontent. In particular, the economic transformation created widespread income insecurity for workers. The autocracy also created discontent among nobles, merchants, and capitalists.

Into this petri dish of discontent was planted the classic germ of uprisings—a food crisis. From 1845, potato crops failed, causing widespread hunger in Europe. When the wheat and rye harvests proved disappointing in 1846, a problem became a crisis.

Three days of turmoil in Paris in 1848 resulted in the overthrow of French king Louis Philippe. Back then, as is the case today, the underlying problems driving the upheaval were common to most European nations, so the French fire quickly became a European firestorm.

By the end of 1848, uprisings had occurred in dozens of nations. But strangely enough, little changed. While tens of thousands died as riots were violently suppressed, few governments changed. The year was, as the English historian Trevelyan put it, "the turning point at which modern history failed to turn."[10] Or, more precisely, history put on the turn signal, but it took European society another century to find the proper turn-off.

The real turning points came in the first decades of the twentieth century— and they took the form of governments, not riots. Karl Polanyi, who coined the term "Great Transformation," viewed communism and fascism as the most revolutionary backlashes against the transformation. To these we should add the election of President Franklin D. Roosevelt with his New Deal economics (known broadly as the social market economy in Europe).

---

10. Quoted in Carl Wittke, "The German Forty-Eighters in America: A Centennial Appraisal." *The American Historical Review* 53, no. 4 (1948): 711–725.

The Great Backlashes: Fascism, Communism, and New Deal Capitalism

At the dawn of the twentieth century, it was plain to all that automation and globalization represented the way of the future—the way to permanently improve the human condition. But the upheavals and backlashes highlighted problems.

Many thinkers viewed laissez-faire capitalism as the wrong way to govern the progress, the wrong way to complete the Great Transformation. Leaving the momentous social and economic choices to capital owners and individual entrepreneurship—guided only by market forces—was the wrong way to harness the promise.

Labor markets were the fundamental issue since people are what society is all about and "labor" is what we call people in an economic setting. The problem lay in three things: average incomes weren't too far from subsistence levels, workers' incomes depended solely on their earnings, and labor was bought and sold like a commodity.

Under these conditions, livelihoods could be won or wasted—all based on the vagaries of faceless market forces. Such fluctuations in supply and demand perpetually exposed large shares of the population to life-threatening uncertainty. In one way, Catherine Spence was essentially killed by a stock-market crash.[11] This unbridled income insecurity, economic fragility, and poverty was not to stand.

A day's work is not a commodity like a sack of wheat—and this is true for one very obvious reason. Labor has recourse to ballots, and if that fails, to bullets. The challenge of fixing the system generated considerable intellectual, social, and political soul-searching.

The basic question was this: How could labor be sheltered from the full force of unfettered markets? The devastation, death, and economic

---

11. According to some, unbridled income insecurity, economic fragility, and poverty didn't seem to be bugs in the system; they seemed to be a feature—a feature those in charge appreciated. According to revolutionary thinkers like Karl Marx, the economic generals of the Industrial Revolution depended on the "industrial reserve army" of unemployed and vulnerable workers to keep the value-creation engine turning smoothly.

dislocation that came with the First World War opened minds to radically new approaches. The early part of the twentieth century tried out three answers: communism, fascism, and New Deal capitalism.

The Communist Manifesto was published in 1848 and thus was part of the historical turning point that history failed to take. But history did take this turning in 1917 in the form of the Russian Revolution. The communist solution was to remove the market from the system entirely.

Society's great choices were not to be made by individuals based on self-interest and guided by the market's invisible hand. They were to be made in the interest of the people and guided by the very visible hand of the Communist Party. The market was out; the plan was in. This would shelter people from the side effects of progress.

The degree of economic control that this implied required absolute political control, so communism soon slipped into a form of dictatorship. Fascism, another radical alternative tried at about the same time, also led to dictatorships.

The Fascist Manifesto was published in 1919.[12] Many at the time viewed fascism as a sensible way of smoothing out the roughest edges of laissez-faire capitalism while avoiding the radical changes of communism. Indeed, for much of the early 1900s, one key justification for supporting fascism was that it was the only real alternative to communism.

The Manifesto called for voting rights for all, including women; proportional representation in parliament; abolition of the wealth-dominated Italian senate; implementation of an eight-hour workday for all workers; and a progressive tax on capital.

Remember that fascism in the 1930s was as yet untarnished by its current association with the horrors of Hitler-ism. The University of Lausanne, for example, awarded the Italian fascist dictator, Benito Mussolini, an honorary doctorate in 1937.

More generally, the fascist response to the backlash against laissez-faire capitalism was to stay with the market for many things but to remove the uncertainty by relying on cooperation instead of competition. Capitalists,

12. In the original Italian it was *Il manifesto dei fasci italiani di combattimento*.

labor, and government would work together for the betterment of all in what was called the "corporatist model." Class conflict was out; class co-operation was in.

Benito Mussolini took power in 1922 and progressively undermined the institutions of democracy to establish a dictatorship. But on the economic front, he was, at first, viewed as a hero of the downtrodden.

He instituted broad welfare spending and public works programs. Swamps were drained to gain farmland, railroads were improved to foster business, and hospitals were built to care for the ill. Fascism, in its early days, was widely admired. It looked even better after the Great Depression brought European and American economies to their knees. Hitler came later, and his national socialism produced some of humankind's greatest horrors. But in its early days, it, like Italian fascism, looked good economically.

Geography and policy shielded the US from much of the turmoil driving European discontent in the early 1900s. This delayed the backlash, but the Great Depression hit Americans hard.

Hunger Marches and FDR's Election

Hunger—which many thought had been banished from advanced industrialized economies decades earlier—returned with the Great Depression's mass unemployment. Not everyone took this sitting down. The Ford Hunger March, organized by the Communist Party USA, was a small but telling example.

On March 7, 1932, a few thousand people marched from Detroit, Michigan, to Ford Motor Company's biggest factory in nearby Dearborn. The goal was to deliver a petition that demanded rehiring of laid-off workers, and the right to organize a labor union. When the protesters reached Dearborn, police attempted to turn them back with tear gas and baton charges. When that failed, police fired into the crowd. Five died.

The protesters' demands were never delivered to Ford, but the event helped to spook the industry into allowing unionization. Better that, industry felt, than the more extreme outcomes that were gaining traction in Europe. There were similar marches in Britain. The year 1932, for example, saw a "National Hunger March" organized by the British Communist party. The aim was to raise awareness of the problem in general by delivering a petition to Parliament that had been signed by a million people.

A hundred thousand marchers showed up. Falling back on a nineteenth-century pattern, the march was violently repressed and the petition confiscated; it never reached Parliament. Protests were seen across the British Isles in the 1930s, especially the areas worst hit by the economic downturn such as Manchester, Birmingham, Cardiff, Coventry, Nottingham, and Belfast. Similar marches as well as mass strikes were common across all the advanced industrial economies. This was a turning point at which history ended up turning.

The Great Depression was launched by a historic stock market crash in 1929 that was made much worse by poor policy. Allowing banks to fail proved deadly, but the real fault went much higher. The sitting president, Herbert Hoover, stuck to his philosophic belief in minimal government. Using workhouse logic that would have made Thomas Malthus proud, he argued that helping the destitute would tempt them into laziness and dependency. As the 1929 recession became the Great Depression, a backlash became inevitable.

In the United States, this took the form of an electoral landslide for a new type of politician—one who promised to end the view of poverty as a moral failing on the part of the poor and who viewed it as the government's duty to be caring and interventionist. Franklin D. Roosevelt, known as FDR, won the 1932 presidential election by 17 percentage points in the popular vote. He took 472 electoral college votes out of 531.

Every backlash ends somehow—usually in some combination of repression and reform. The question of whether the repression and reform represent a resolution is something that can only be answered by history. As it turns out, both communism and FDR's policies were lasting resolutions to the core shortcomings of nineteenth-century capitalism.

BACKLASH PRODUCES RESOLUTION

Roosevelt's radical changes, called the "New Deal," rested on the "3Rs": "relief" for the poor and jobless; "recovery" of economic activity to pre-crisis levels; and "reform" of the economy to eliminate the causes of the economic collapse, and social and economic despair.

Key reforms included pro-labor union laws, higher income taxes on the rich, and thorough regulation of banks and anti-competitive practices. Workers' economic vulnerability was massively reduced since big business now had to negotiate with big labor. New Deal programs also directly supported disadvantaged groups ranging from farmers and the unemployed to youth and the elderly. Since the changes came via a democratic election, the radical solutions catching on in Europe at the time failed to catch the fancy of the American working class.

Under Roosevelt, US government spending jumped from about 5 percent of national income to about 20 percent—where it has stayed ever since. The WWII military spending receded and was replaced by New Deal spending, especially on pensions and healthcare.

FDR was president for twelve critical years—from 1933 to 1945. His successors changed little. Even Republicans like Eisenhower and Nixon accepted FDR's basics, and the New Deal was expanded by President Lyndon Johnson in the 1960s via his Great Society program.

Similar economic programs were adopted by governments in all the Western, industrial countries. The key shift was a tectonic realignment of responsibilities. All around the world, governments took responsibility for social justice and the plight of the disadvantaged. Henceforth, markets were viewed as being in charge of economic efficiency; governments were viewed as being in charge of social justice.

Fascism was ended by force of arms in the 1940s. Communism and New Deal capitalism both flourished—giving rise to a fifty-year struggle between them. Even after hardline communism was widely discredited by the fall of the USSR, it continued to thrive in a massively reformed form. Today, a heavily modified, market-friendly version of communism rules in China and a few other nations like Vietnam and Cuba. In essence,

communism only survived by becoming more like capitalism while capitalism survived only by become more like communism.

The various resolutions of the backlash in the 1920s and 1930s set the
modern world on a steady course for decades. The fruits of social calm,
booming innovation, and advancing globalization yielded what the French
call *les trente glorieuses*.

Thirty Glorious Years

Once Roosevelt's New Deal reforms made the whole socio-economic
system politically sustainable in the United States, and similar reforms
did the same in other industrial nations, economic growth boomed in the
West (as the capitalist world came to be called despite including Japan,
Australia, and New Zealand).

For decades, postwar innovation, automation, and globalization
produced the fastest income growth the world had ever seen—twice as
fast as Great Transition growth. But the innovations did far more than
accelerate incomes. The new innovations produced a massive reduction
in income inequality and generalized prosperity and economic security.

These innovations mostly concerned the making of things, including
lots of new things. The inventions were, in short, a gargantuan pull factor
into industry. Best yet from the social stability perspective, the rising
number of high paying manufacturing jobs were for people with average
skill levels. These were jobs that required some thinking and some perception skills—things that machines couldn't do—but nothing that required
advanced education or remarkable dexterity.

The result was the emergence of a great middle class—people who
owned homes and cars, had good jobs, and formed stable communities.
The income distribution was massively compressed to the extent that few
felt that the rich were getting rich because the poor were getting poorer.
President Kennedy could claim, in 1963, that "a rising tide lifts all boats,"
and he was right. The thirty years after the war were simply an economic
miracle. All you needed to do well in those thirty glorious years was a high
school degree and a willingness to work—or so it seemed to many.

The "ground zero" of these innovations was manufacturing. The special-century inventions were clearly most favorable to people who made things in factories, but the innovations also helped workers in the service sector. The inventions—by ushering in the modern era—boosted the productivity and living standards of nearly everyone.

Workers involved in utilities, transportation, cleaning, and wholesaling and retailing found it easier to do their jobs with motor vehicles and electric power tools. The progress also made professionals, like lawyers, doctors, architects, and engineers, more effective in ways ranging from electric lighting, air conditioning, and X-ray machines, to home appliances, ballpoint pens, typewriters, and carbon paper.

Value creation and capture still lay in the hands of firm owners—capitalists, if you will—yet the New Deal reforms improved the social outcome. Strong labor made sure industry shared the fruits of productivity gains with the workers. Monopolies were subject to tight scrutiny, and businesses had to respect health, safety, and environmental regulations. Government subsidized education and established excellent public universities where people could earn advanced degrees at affordable prices.

LESSONS, MECHANISMS, AND THE NEXT TRANSFORMATION

The Great Transformation started with a mighty technological impulse that launched a four-step progression: transformation, upheaval, backlash, and resolution. The tech impulse triggered the economic transformation by unleashing the disruptive duo of automation and globalization, but not both at once. It first triggered mechanization, or what today we call automation. The result was a virtuous, self-reinforcing cycle of innovation, industrialization, and rising incomes.

A century later, the technology impulse triggered globalization. Once the tech-trade team was in the game, the happy helix driving economic transformation was accelerated by innovation-led growth.

While this was a good thing overall, the dynamic duo of automation and globalization transformed the economy in ways that produced

wonders and woes. The transformation disordered people's lives
along with the whole traditional economic architecture of value creation and capture. The changes upset communities, altered lives, and
created triumphs and tragedies. The pain-gain package, in short,
produced economic, social, and political upheaval. The upheaval placed
intolerable strain on the social, economic, and political fabric of the time.
The changes came faster than societies could adjust to them. And, since—
as the old saying goes—things that can't go on, don't, they didn't. The final
of the four steps was resolution. Two of the three solutions—communism and New Deal capitalism—are still with us. The third, fascism, was
extinguished by the main adherents of the other two.

Another lesson from the Great Transformation concerns jobs displacement and job replacement—topics that are at the heart of today's "future
of work" deliberations.

Automation and globalization drove a sensational re-orientation of the
economy. Taking Britain as an example, the share of workers in industry
rose progressively from 19 percent in 1700 to 49 percent in 1870, according
to one of the grand masters of economic history, Nicholas Crafts.[13] During
this period, the nation also shifted from a primarily rural society to one
where almost two-thirds of people lived in urban areas. Much perspective
can be gained by taking a closer look at the jobs shift.

Open versus Sheltered Sectors

During the Great Transformation, as is true today, the disruptive duo—
automation and globalization—didn't touch all sectors of the economy
equally. Some sectors were open to the disruptive duo's influence, while
others were sheltered from it. This uneven impact of automation and globalization across sectors goes a very long way to explaining the historic
shifts in jobs from farm to factory. And it helps us understand the impact

---

13. Nicholas Crafts, "British Industrialization in an International Context," *Journal of
Interdisciplinary History* 19 (1989): 415–428.

of past, present, and future automation and globalization. The basic notion is uncomplicated.

Sheltered sectors tended to gain jobs since displaced workers had to, and in fact did, go somewhere. Or more precisely, over the medium term, wages adjusted to the point where it became worthwhile to create jobs for most people. Services in the Great Transformation were shielded from globalization since most services require face-to-face interaction. Quite simply, you can't put services on a steamship the way you can with grain and textiles. Service jobs were also largely shielded from automation since the technological impulse focused on helping people make things, not think about things.

The new service jobs were wide ranging and often linked to higher incomes. The rise of the middle class meant that there were many people with cash left over after paying for food, housing, and clothing, and they spent some of the cash on services that made their lives better and easier. For open sectors, things were subtler.

Sectors that were most directly open to automation could see rising or falling employment depending upon magnitudes—depending upon which side won the productivity-production foot race.

Structural Transformation

Taking Britain as an example, the left panel of Figure 2.3 shows the number of jobs in the three major areas—services, manufacturing, and agriculture. The right panel shows the same numbers as a share of jobs.[14]

One striking feature that can be seen by comparing the two panels is how the absolute number of jobs rose in all sectors up till the mid-1800s, even if jobs in manufacturing rose faster. The reason was the booming UK population growth and the fact that markets and entrepreneurship

14. For details and data see Berthold Herrendorf, Richard Rogerson, and Ákos Valentinyi, "Growth and Structural Transformation," Chapter 6, in Philippe Aghion and Steven Durlauf (eds.), *Handbook of Economic Growth*, vol. 2B (Amsterdam and New York: North Holland, 2014).

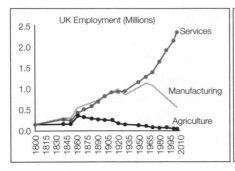 

Figure 2.3 Structural Transformation: UK Employment Pattern, 1880–2008.
SOURCE: Author's elaboration of data published in Berthold Herrendorf, Richard
Rogerson, and Ákos Valentinyi, *Handbook of Economic Growth*, vol. 2B, ch. 6, "Growth
and Structural Transformation," http://dx.doi.org/10.1016/B978-0-444-53540-5.00006-9.

eventually found something for everyone to do. The absolute decline in
agricultural employment came later.

A second feature to note is the way that the sheltered service sector expanded in line with the open manufacturing sector until the 1970s. The
sheltered service sector was a natural absorber of many of the workers
entering the rapidly expanding workforce.

The Great Transformation pattern for the US is similar, but it starts
with a far higher share of workers in farming and a far lower share in
industry—at least in part because imperial Britain suppressed industry
in its colonies. While there are substantial differences in the two Great
Transformation patterns, these are largely down to initial conditions, and
the rather special nature of the US—especially its expanding landmass.

In America, employment in all three sectors rose rapidly until the early
1900s. Just as in England, the dynamic duo of trade and mechanization
was creating millions of new jobs in industry, and rising incomes were
creating millions of service sector jobs. The introduction of railroads, ac-
quisition of new land, and the construction of inland waterways had the
effect of grandly expanding the amount of arable land. That, plus mass
migration from Europe, resulted in booming farm-sector employment.

The shares shown in the right panel of Figure 2.4 display the classic
structural transformation of an agrarian/rural economy into an urban/

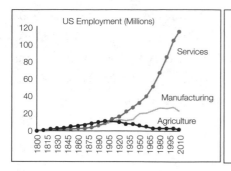

Figure 2.4 Structural Transformation: US Employment Pattern, 1880–2008.
SOURCE: Author's elaboration of data published in Berthold Herrendorf, Richard Rogerson, and Ákos Valentinyi, *Handbook of Economic Growth*, vol. 2B, ch. 6, "Growth and Structural Transformation," http://dx.doi.org/10.1016/B978-0-444-53540-5.00006-9.

industrial one. Agriculture's share plummeted, while services and manufacturing shares soared. The number of US jobs in manufacturing rose for much longer than in the UK—even though the two nations' share figures fell from about 1965. The driving forces behind the differences were mostly population growth and the fact that most manufacturing was sold domestically, so a big population meant a big customer base. The US population rose by about 125 million between 1850 and 1950, while the UK's rose by only 27 million. And the rapid US expansion continued. In the two decades following 1950, the number of Americans increased by 20 million, while the number of Brits increased only by 5 million.

As both sets of charts illustrate, something historic changed at the end of *les trente glorieuses*. The steady shift in the share of workers in industry turned on its head.

THE SERVICES TRANSFORMATION

Catherine Spence's demise in the London Docklands started our account of the Great Transformation. The demise of the Docklands itself ends it. For centuries, the Docklands rolled their way through booms, busts, and bombings—becoming the Royal Docks in the process. The killing blow came when shipping technology rendered the Docklands uncompetitive with deep-water ports further down the Thames. At the end of the 1970s, the docks were shuttered. The area was left to weeds, wildlife, and winos.

The transformation of the Docklands is a handy symbol for the second great economic transformation that started in the 1970s. This great economic transformation switched advanced economies from industrial to post-industrial—to places where most workers worked in offices, not factories or farms.

But why the change?

3

The Second Great Transformation: From Things to Thoughts

"The present administration . . . has either forgotten or it does not want to remember the infantry of our economic army . . . the forgotten man at the bottom of the economic pyramid." Franklin D. Roosevelt spoke these words in the deepest depths of the Great Depression.

In 2017, another populist politician said: "The forgotten men and women of our country will be forgotten no longer." That was President Donald Trump, who was elected in a backlash against an economy that, for decades, provided more wealth for the well-off but more anguish for the average. Since the 1970s, the US working class has seen stagnating wages, rising economic insecurity, and increasing hopelessness. The situations in Europe and Japan are not as dire, but they share the trends.

FDR's reforms fixed American capitalism and set the stage for the thirty glorious years of economic prosperity. So why are we back here again? Why aren't the disruptive duo of automation and globalization lifting all boats? Why has the tech-trade team flipped from the factory-job creating force it used to be after World War II to the factory-job destroying force it is today?

The answer is as simple as it is strange.

A Very Different Technological Impulse—Helping Brains,
Replacing Brawn

A new technological impulse kicked in when computers and information
technology became practicable. The new technology produced a new type
of automation in the early 1970s, and—twenty years later—a new type of
globalization. This new "tech-trade team" plays by a very different set of
rules than the last one did.

The new technology provides better tools for those who work with
their heads, but better replacements for those who work with their hands.
The new technology—Information and Communication Technology
(ICT)—focuses on intangibles, not tangibles. It is all about processing,
transmitting, and storing information. This difference matters.

Post–World War II prosperity was driven by a technology that favored
the making of things. The resulting automation-globalization duo directly
boosted the productivity of people who worked with their hands. It helped
people who worked with their heads, but only indirectly because it was a
technology of things, not thoughts. It created masses of new industrial
jobs. Even better, since most people back then worked with their hands,
the more-manual-than-mental aspect of the tech-trade team did wonders
for social cohesion.

The 1970s technological impulse did just the opposite.

Creating better replacements for factory workers—robots and the
like—was a massive push factor that emptied factories faster than the
Great Transformation emptied farms. The better tools for brain workers,
by contrast, was a massive pull factor for office workers and professionals.
It created millions of new service-sector and professional jobs—many of
them in occupations that were previously unimaginable.

From the social cohesion point of view, the new technology was divisive. Since the "head workers" were already better off than the "hand
workers," a technology which favored brains over brawn favored the few
who were already favored, while disfavoring the many who weren't.

The London Docklands once again provides the perfect portrait.

Canary in the Docklands

From the year Catherine Spence starved to death and right up to the 1970s, the London Docklands were the gateway for goods coming into Britain and goods going out. The docks were all about things, not thoughts. And they provided thousands of good working-class jobs directly, and tens of thousands more indirectly.

That ended when the last commercial vessel was unloaded on December 7, 1981. The closure of the Docklands created economic and social problems. Although no one starved as in 1869, local unemployment rocketed, crime rose, and social ills multiplied. Today, however, the area is booming—especially one development called Canary Wharf.

The goods-based economy has been completely replaced by an information-based economy. Carney Wharf is now one of the most important financial districts in the world. In the boom years running up to the financial crash, a single building sold for a billion dollars. Not bad for an area that had, a few decades earlier, been left to weeds, wildlife, and winos. But while the Docklands are now posh and pulsing with economic activity, it is definitely not lifting all boats.

Highly educated workers who earn astronomical salaries dominate the place. The area employs plenty of unskilled workers pulling lattes, pushing brooms, and shining shoes, but there are precious few jobs to support a prosperous middle class. The Docklands is now an industry of thoughts, not things.

This new phase of structural transformation is called the post-industrial transformation, but it is really a second great transformation, call it the Services Transformation.[1]

---

1. Some call this the third industrial revolution, even though it is mostly about deindustrialization and the rise of services. See Jeremy Rifkin, *Third Industrial Revolution: How Lateral Power Is Transforming Energy, the Economy, and the World* (New York: St. Martin's Griffin, 2011).

New Technological Impulse, New Four-Step Progression

The new ICT impulse launched a second great transformation and a second four-step progression (economic transformation, upheaval, backlash, and resolution). This new economic transformation was not as great as the original Great Transformation, but it did disorder the lives of millions and reshape economic social and economic realities into what the sociologist Alain Touraine called the "post-industrial society."[2] Jobs shifted from factories to offices, urbanization continued, many rural communities declined or disappeared, and the fulcrum of value creation shifted from capital to knowledge. The nature of globalization changed, and the unquestioned economic dominance of the West was questioned by facts on the ground.

This economic transformation produced upheaval—just as it did in the nineteenth century. The twenty-first-century upheaval was nowhere near as great as that of the nineteenth and early twentieth century. It was, nevertheless, traumatic—especially in the US where government safety nets had been removed or not put in place as they were in Europe and Japan.

The social and economic upheaval produced a backlash in 2016 with the Brexit vote and President Trump's election. This was far more moderate than what we saw in the early 1900s, but when it came, it shattered realities. It continues to shake the global order. And resolution has yet to come.

Was 2016, like 1848, a turning point where history failed to turn? Was 2016 just one small backlash, like the Luddites, that will eventually lead to a large backlash on the order of fascism, communism, or New Deal capitalism?

There can be no clear answer to these critical questions since the future is unknowable. But the future is also inevitable, so it is best to start at

---

2. Alain Touraine, *The Post-Industrial Society. Tomorrow's Social History: Classes, Conflicts and Culture in the Programmed Society* (New York: Random House, 1971).

the start and identify trends that will guardrail our thinking about future developments.

We start with the technology. As with steam, it took a while to work the bugs out.

NEW TECHNOLOGICAL IMPULSE

The Hamtramck auto factory in Detroit, Michigan, was supposed to be "the most modern auto plant in the world," according to General Motors (GM) chief Roger Smith. But that's not what he was calling it after they turned on the lights and ramped up production in 1985.

What was supposed to be a showcase for the cost-cutting and quality-boosting advantages of industrial robots turned into a clump of chaos. The painting robots melted the plastic taillights and occasionally went wild, painting each other, and the walls as well as the cars. The robots fitting the windshields sometimes got confused and sent the glass smashing into the car instead of installing it gently. Other robot confusions led to Buick bumpers being fitted onto Cadillacs. The computer-controlled vehicles delivering parts to the line sometimes froze.

As Thomas Bonsall puts it in his book, *The Cadillac Story: The Postwar Years*, "Many of the extravagantly expensive devices did not work at all— which may have been a blessing considering the mayhem caused by the ones that did." Maybe it was sabotage, or just an example of the old saying, "To err is human, to really foul things up requires a computer." The foul-up took years to fix. But fix it they did.

Hamtramck was a mere speed bump on the way to replacing autoworkers with automatons. Automation has been replacing US and European factory workers ever since. Computers, as it turned out, were driving a very different kind of automation than the special-century technologies did during the thirty glorious years.

A Technological "Continental Divide"

When computers and integrated circuits started getting useful in the 1970s, automation crossed a "continental divide," as mentioned. Most machines before this divide were either rigidly devoted to a single task, or required a human to direct them. The famous seed drill of Jethro Tull (the eighteenth-century inventor, not the twentieth-century rock band), for example, was a complicated contraption that could do only one thing. It carved three rows into the dirt, dropped seeds into these at specific intervals, and then covered them with the right amount of soil. Other machinery—say, a press drill—could do lots of different things, but it required a human brain to make it useful. ICT disrupted this pattern by making machinery more flexible without human brains.

An early version was "numerical controlled machines." These were generic machines—lathes, drills, and the like—that were controlled by a program that could be changed to deal with different jobs. At first, the controlling instructions were fed in using a one-inch-wide punched tape. A "controller unit"—a sort of computer—read and interpreted the instructions and converted them into mechanical motions by the machine tool.

The newfound flexibility of machine tools destroyed one part of humans' comparative advantage in factories—namely, their ability to learn new tasks, adapt to evolving situations, and react flexibly.

The 1973 Milestone

Dating exactly when the continental divide was crossed is difficult since progress is a process, not an event. Nonetheless, 1973 is a convenient starting date since it is the year that Texas Instrument employees Gary Boone and Michael Cochran patented the first "computer on a chip." This was revolutionary.

Putting a computer on a chip made earlier approaches to building computers obsolete; before 1973, computers were built up from racks of circuit boards. By combining on a single thumbnail-sized device the "brain" (central processing unit, or CPU), digital memory, and circuits to handle inputs and outputs, the computer-on-a-chip reduced the cost and improved reliability—all while reducing power usage and thus solving overheating problems. Soon, industry was having chips with everything.

By sticking a computer-on-a-chip into a robot arm, many repetitive mechanical tasks could be automated and the same robot could be quickly reprogrammed to do other tasks when the time came.

In terms of globalization, plummeting communication costs had an effect on the world economy akin to the impact of steam power. In particular, the cost savings revolutionized manufacturing. Before ICT, most stages of production had to be placed within walking distance in order to coordinate the complex processes. Just as steam power made it economical to separate production and consumption over long distances, the communication part of ICT allowed companies to place some stages of production abroad.

The new ICT impulse produced a new economic transformation, as I point out at length in my 2016 book, *The Great Convergence: Information Technology and the New Globalization.*[3] The societal changes weren't anywhere near as epic as those of the Great Transformation, but they still shook things up in a big way. Industrialization—which had been the codeword for progress for a couple of centuries—turned into deindustrialization. The results were dramatic.

NEW TECHNOLOGY PRODUCES A NEW ECONOMIC TRANSFORMATION

The impact of the ICT impulse was first felt though the automation of industrial jobs. Computer-controlled machines rapidly displaced workers, especially in the auto industry, and especially those involved in welding, painting, and specific pick-and-place tasks. As ICT advanced, the

3. Richard Baldwin, *The Great Convergence: Information Technology and the New Globalization* (Cambridge, MA: Harvard University Press, 2016).

repetitive, manual tasks that industrial robots could handle increased—displacing jobs as it went.

From the 1990s, many factories in advanced economies turned into computer systems where the peripherals were industrial robots, computerized machine tools, guided vehicles, and so on. Roger Smith's dream of Hamtramck-like factories supplanting workers came true, or mostly true. Factories became places where workers helped machines make things, not the other way around.

The impact on factory employment was dramatic.

The new technological impulse has been a massive and sustained push factor—pushing workers out of manufacturing in advanced economies. In all advanced economies, the share of jobs in manufacturing has been on a "mission to zero" since the 1970s, as Figure 3.1 shows. Manufacturing employment shares in the United States fell from 30 percent in the 1970s to something like 10 percent in the 2010s. The United Kingdom's industrial sector, which used to absorb over a third of workers, now accounts for only one in ten jobs. The manufacturing share in Germany halved from 40 percent to 20 percent, and Japan's declined from 27 percent to 17 percent.

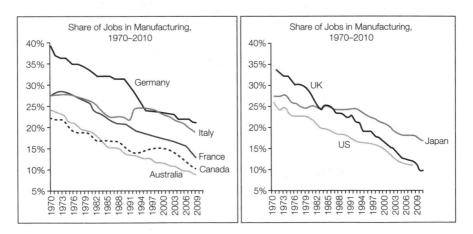

Figure 3.1 Share of Manufacturing Jobs in Advanced Economies, 1970–2010.
SOURCE: Author's elaboration of UNSTAT online data.

The new technology was, by contrast, a pull factor for office workers and professionals. Occupations in which people worked more with their heads than their hands found that ICT made them more productive. It created radically more efficient processes for doing all sorts of service-sector tasks. When I was an intern at the US Senate's Joint Economic Committee in Washington in the summer of 1979, I wrote a research paper, and that meant writing it out in longhand (try it one day and you'll understand where the "long" in longhand comes from). A typist typed it. In 1991, when I worked as an economist at the Council of Economic Advisors in the Bush (senior) White House, I wrote everything on a PC and printed it out. That made everything faster, even though sending it to people had to be done by post or by hand (the government didn't have email back then).

The ease of gathering and manipulating data lowered the price of many services, like design and editing services, and this greatly boosted their consumption. It also led to many new products in the service sector. Software became an industry. Telecommunication introduced all sort of new services and e-commerce was invented. Millions of new service-sector jobs were created as the service-sector expansion mirrored the continued decline of farm and factory jobs.

While the first couple decades of ICT had enormous impact on automation, from 1990 or so, ICT came to have enormous effects on globalization. But this globalization was not like the one that started in the 1800s and dominated all through the thirty glorious years. A new kind of technological impulse resulted in a new kind of globalization.

What Puts the "New" in the New Globalization?

Since the dawn of civilization, high cost of moving goods, ideas, and people formed a "glue" that bound production to consumption geographically. People were bound to the land on which they grew their food, and production was bound to the people. Each village was largely self-sufficient in everything from food and footwear to tools and textiles. This was before the Great Transformation.

As technology advanced, all three costs fell—but not all at once. The first technological impulse—steam power—radically reduced transportation costs. This ended the need to make goods close to where they were consumed. Once this change made long-distance trade feasible, the huge price differences across the world made it profitable. Trade in goods boomed from the early 1800s as the steam impulse was augmented by later developments like steel hulls, diesel engines, containerized cargo ships, air cargo, and worldwide trade liberalization. These advances lowered the cost of moving ideas and people as well, but not in a revolutionary way.

Strangely enough, as production dispersed across nations in this first phase of globalization, it clustered within nations into factories and industrial districts. This microclustering wasn't done to save trade costs; it was done to save on communication costs—namely, the cost of moving ideas. The point is that being able to sell to the whole world favored large-scale, highly complex production processes. To manage the complexity, firms moved all the production into one place. Stages of production, in other words, bundled into factories.

ICT lowered the cost of moving ideas even faster than steam had lowered the cost of moving goods. This, in turn, ended the necessity of performing most manufacturing stages inside the same factory or industrial district. The improved communications that came with the ICT revolution had mammoth implications for the spatial organization of factories—what came to be called "offshoring." The manufacturing microclusters— factories and industrial districts—that were so prominent up to the 1980s had been held into these tight clusters by the high cost of long-distance communications much more than the high cost of transportation.

American companies had long understood that they could perform some aspects of the manufacturing process more cheaply abroad. The highly modular nature of the semiconductor production process, for example, allowed US semiconductor producers to put some stages in Asia as early as the 1970s.[4] The barrier to doing this in most industrial sectors

4. Jeffrey W. Henderson, *The Globalization of High Technology Production* (New York: Routledge, 1989).

was the high costs of coordinating production. That's why offshoring only really started racing after ICT made international coordination cheap and reliable. Only then could companies in the United States, Germany, and Japan unbundle complex production processes geographically without much loss in quality, timeliness, or reliability.

This new possibility created the new globalization. It allowed manufacturing firms in advanced economies to exploit the vast international wage differences between, for example, the United States, Germany, and Japan on one hand, and nearby developing nations like Mexico, Poland, and China on the other hand. The result was a quite sudden and massive deindustrialization of the advanced economies.

In 1970, the advanced industrial economies known as the G7 (United States, Japan, Germany, Britain, France, Italy, and Canada) produced over 70 percent of the world's manufactured goods. That declined gently during the 1970s and 1980s, but from 1990 it plummeted. The G7 share fell from two-thirds to less than half in just twenty years, as Figure 3.2 shows.

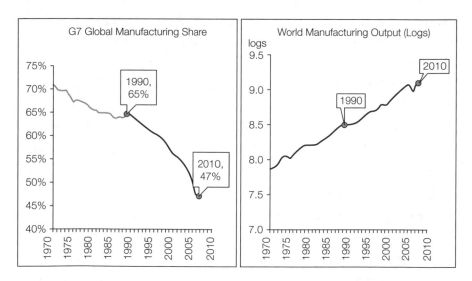

Figure 3.2 G7 Global Manufacturing Share and Global Manufacturing Growth, 1970–2010.
SOURCE: Author's elaboration of BLS online data.

The chart also shows that nothing radical happened to the overall growth in world manufacturing output. Putting together these two puzzle pieces tells us that G7 manufacturing went somewhere else. That "somewhere" was the emerging economies, primarily China.

This was one of the most dramatic aspects of the Services Transformation. The historically fast deindustrialization of the former industrial giants, and the historically fast industrialization of a handful of formerly unindustrialized economies—call them the Industrializing 6 (China, India, Indonesia, Korea, Poland, Thailand, and Turkey). Most economists misthink this massive flip in the world of manufacturing by focusing on the production that was offshored. In reality, it was about thoughts, not things.

As I detail at length in my 2016 book, *The Great Convergence: Information Technology and the New Globalization*, knowledge is the key to understanding this rapid deindustrialization. The point is that the US, German, and Japanese offshoring firms sent along their know-how with the offshored stages of production and displaced jobs. How could they have done otherwise?

When Toyota makes parts in China for inclusion in the cars they assemble in Japan, the company can't rely on Chinese technology. Instead, Toyota sends its know-how to China to ensure that the Chinese workers are doing the right thing and in the right way. As a result, the flows of knowledge that used to happen only inside Japanese factories became part of international commerce.

It was exactly these new technology flows that triggered the rapid industrialization in China and a few other developing nations. It started with production directed by multinationals, but domestic production boomed as the know-how diffused more widely.

The thing that puts the "new" in the new globalization is the technology that started crossing borders from 1990 or so. Offshoring did lead to more trade in parts and components, but that wasn't the revolutionary part. The thing that changed the world was the colossal, one-way flow of technology from mature to emerging economies. This is a really key point, so a bit of elaboration is in order.

A football analogy helps clarify. "Imagine two soccer clubs sit down to discuss an exchange of players. If a trade actually occurs, both teams will gain. Each team exchanges players of a type they had too many of for players of a type they had too few of. Now consider a subtly different type of exchange. Suppose on the weekends, the coach of the better team goes to the home pitch of the worse team and starts to train their players."[5] The exchange of players is like the old globalization—goods crossing borders. The coaches training is like the new globalization—know-how moving in one direction.

These new knowledge flows spawned a new reality in manufacturing globally.

Before this widespread offshoring on manufacturing jobs, international competition in goods was based on one of two choices. Firms in developing nations could rely on low technology and hope that their low wage more than compensated for the technical inefficiency. Firms in advanced economies, by contrast, used high technology and hoped this would more than compensate for the high wages they had to pay advanced economy workers.

From about 1990, a third way opened. Manufactured goods could be made with high technology that had been offshored to low-wage nations. This transformed the world of manufacturing. It explains why the Industrializing 6 industrialized so rapidly. They didn't have to develop the technology themselves. The offshoring companies brought everything needed except the labor. You could call it "add-labor-and-stir" industrialization. And this is not as obviously a win-win outcome as was the old globalization.

The rapid industrialization of the Industrializing 6 was surely good for them. It is not at all sure that advanced economy factory workers also benefited. American, European, and Japanese workers no longer had privileged access to the know-how developed by their national firms. The monopoly that advanced-economy workers used to have on

---

5. This is from the introductory chapter in Jeffrey Baldwin, *The Great Convergence: Information Technology and the New Globalization* (Cambridge, MA: Belknap, 2016).

advanced-economy technology was broken. American, German, and Japanese companies taught foreign workers to make parts and components that used to be made domestically; this teaching hastened the loss of factory jobs in G7 nations.

In a nutshell, it was knowledge that changed globalization and ICT that allowed the knowledge to flow. The new know-how flows also explain the very different impact of the new globalization.

The New Globalization's Very Different Economic Impact

There are four differences between the old and new globalization that stand out. Globalization's impact became more individual, more sudden, more uncontrollable, and more unpredictable.

It was more individual since it didn't just happen at the level of sectors and skill groups. Globalization during the Great Transformation was felt at the level of sectors—say, semiconductors, or earthmoving equipment. This was true since foreign competition showed up in the form of products that were made in particular sectors. Moreover, since some types of labor—say, unskilled labor—were more important in some sectors than others, globalization's impact tended to fall unevenly on skill groups. In the postwar period, for example, globalization tended to help skilled workers and hurt unskilled workers.

With the New Globalization, the extra competition and opportunites can help or hurt workers in one stage of production while helping workers in other stages in the same firm. To put it differently, the new globalization operated with a finer degree of resolution. It created winners and losers as before, but they weren't as clearly lined up with winning and losing sectors, or winning and losing skill groups. The new opportunities and competition were more individual. And then there was the speed of the thing.

Before the ICT revolution, globalization transformed societies but slowly. The "change-clock" ticked decade by decade. Since the ICT revolution, the change-clock ticked year by year. Industrialization took a century to build up in the advanced economies. Deindustrialization and the shift of manufacturing to emerging nations took only two decades. The reason for the unprecedented speed was the unprecedented nature of globalization. The emerging markets were not industrializing the way the G7 nations had in the twentieth century. Much of the emerging-market manufacturing take-off, especially in the beginning, was coordinated by G7 firms.

Another defining feature of this new globalization was that it was less controllable. Governments had lots of tools for monitoring the passage of goods and people across borders but very few tools for controlling firms' knowledge crossing them. And since it was the advance of ICT that drove this new globalization, governments had few practicable tools for controlling the pace.

Lastly, new globalization was more unpredictable. Since the 1990s, it has been hard to know which stages of a manufacturing process will be offshored next. This changed nature of globalization created a generalized sense of vulnerability in advanced economies. No one in the manufacturing sector could really be sure that their job wouldn't be next.

As if these shocks weren't enough, the whole deindustrialization phase coincided with a massive, worldwide slowdown in growth.

The Post-1973 Growth Slowdown

Most wealthy nations experienced a slower income growth rate at the start of the second great transformation. Each decade since the 1960s has seen slower per-capita income rises. The decline was gentle but significant in the last three decades of the twentieth century. The drop-off has been much more marked in the twenty-first century. On average, US incomes rose by 3.3 percent per year in the 1960s, but by less than half that in the new century; the figures for the United Kingdom are quite similar. For Germany, the 1960s were a miracle, with growth of almost 4 percent annually, but since 2000, the average has been more like 1 percent per year.

Change is always easier when incomes are, on average, rising quickly. The opposite is true as well. The whole adjustment process was made more difficult by the fact that economic growth slipped into low gear.

The economics profession still does not have a full explanation for this, but one notion that fits tightly into the Serivces Transformation is the story told by Robert Gordon, whose ideas we encountered in Chapter 2. He argues that growth and innovation didn't slow down from the 1970s but rather that they returned to historical norms.

The cluster of new inventions that arose from about 1870 accelerated innovation and thus incomes, but not forever. The collection of new inventions—everything from electric motors to plastics—proved to be a rich pallet with which clever inventors "painted" new products and new ways of making old products. The elements where combined and recombined and the result was decades of above-normal rates of inventiveness and thus above-normal growth.

By the 1970s, according to this theory, the world had developed the bulk of all the new products and processes that were made possible by the special-century techniques. After that, per-capita growth returned to its normal pace of around 1 or 2 percent per year.

The pains and gains that came with the growth slowdown and the new forms of automation and globalization disordered many traditional arrangements. Everything was made more difficult by the slowing of growth. Together, these aspects of the economic transformation caused massive disruption to manufacturing workers and their communities. The result was upheaval.

One fact is critical to understanding the upheaval. The new globalization hit the same workers whose livelihoods had also been hit by the new automation. Manufacturing workers in the United States, Canada, Europe, and Japan found themselves competing with robots at home and with China abroad.

This economic transformation drove an upheaval. One of the most stunning aspects of the upheaval came from what has been called the "skill twist."

NEW TRANSFORMATION PRODUCES A NEW UPHEAVAL

The computer-on-a-chip breakthrough launched a phase when technology made unskilled factory workers more replaceable, while making highly skilled office workers more productive. Economists have recently called this "skill-biased technical change." A livelier term was used in a 1983 study on the employment implications of automation. That report called it the "skill twist."

The 1983 report phrased it this way: "If there is an increase in unemployment as a result of the spread of robotics technology, we fear the burden will fall on the less experienced, less well-educated part of our labor force. . . . The jobs eliminated are semi-skilled or unskilled, while the jobs created require significant technical background." [6]

This is exactly the aspect of the trend that proved so disruptive to the industrial working class in advanced economies. Gone were the days when a high school education and a union card would get you a house in the suburbs with a car in the garage and a pension in the bank. Social problems were magnified as US union power plummeted along with union membership, and the government failed to step up with sufficiently robust retraining schemes.

Factories still needed workers, but the skill twist meant that they tended to be at the extremes of the skill range. High-skilled workers were needed to mind the robots and computers. And unskilled workers were needed to clean the place and handle unexpected manual tasks, but jobs were scarce for those in between. The masses of production line workers were increasingly out of luck.

The result came to be known as the "hollowing out" of the American, European, and Japanese labor markets. Workers at the high and low ends of the skill scale did OK; those in the middle did not.

Meanwhile, the same technology cut out broad swaths of middle-skilled office workers who had been employed to facilitate the gathering,

---

6. H. Allen Hunt and Timothy L. Hunt, *Human Resource Implications of Robotics* (Kalamazoo, MI: W.E. Upjohn Institute for Employment Research, 1983).

processing, and transmission of information. Typists, file clerks, telephone operators, and secretaries were phased out. By contrast, the ICT advances amplified the productivity of college-educated workers who worked with ideas and information.

As in the Great Transformation, the changes weren't just about people changing jobs. There was also a deep movement in who captured the value created. During the Great Transformation, the linchpin factor of production swung from land to capital. In the second great transformation, it swung from capital to knowledge.

A Sea Change in Value Creation and Capture

Capital is not dead, but it's ailing—a point made forcefully by the 2017 book *Capitalism without Capital: The Rise of the Intangible Economy.*[7] Capital has lost the race for supremacy. The book's authors argue that this is nothing short of a "quiet revolution." Today, companies invest more in intangible assets—things like design, branding, patents, R&D, and software—than in traditional, tangible assets—things like machinery, buildings, and computers. Thoughts, not things, if you will.

The sea change started in the 1970s. Investment in tangible assets—let's just call it capital—as a share of the economy peaked around 1979 and has fallen since. Investment in intangible assets—call it "knowledge"—has instead risen steadily. Knowledge overtook capital around 1990.

Increasingly, value is created by labor working with knowledge—either knowledge clusters controlled by firms like Google and Apple, or knowledge stuck into people's heads in the form of education and experience. Increasingly, to control a bit of knowledge is to control the value creation, and thus the value capture. Perhaps we should stop talking about capitalism and start talking about "knowledge-ism." Be that as it may, the shift has transformed our economies.

7. Jonathan Haskel and Stian Westlake, *Capitalism without Capital: The Rise of the Intangible Economy* (Princeton, NJ: Princeton University Press, 2017).

Labor that lacks knowledge is abundant, and although knowledge capital isn't really fixed, knowledge capital owners are increasingly the ones with the power to decide the division of the value created. The average worker has not benefited.

From 1973 to today, the output per hour worked in America rose by over 70 percent. But the fruits of this faster value creation have not been shared. The hourly pay of the average American has risen by about 10 percent, but a gigantic gap has opened between pay and productivity; the value created per hour worked rose steadily, but the average pay of the people doing the work did not rise. Since the value created had to go somewhere—value capture shares have to add up to 100 percent—the question is: Who got the value? The answer is: knowledge owners.

The decades following the 1970s have been a veritable land of milk and honey for those with lots of knowledge in their heads. Americans with higher education have seen their incomes soar. As MIT economist David Autor has shown, the inflation-adjusted earnings of US men with a first university degree or higher rose about 50 percent from 1970 to 2010.[8] Men with some college but no degree saw their wages stagnate over these years. American men with high school educations actually lost ground. They make less today (in inflation adjusted terms) than they did in 1973. For US high school graduates, earnings per week fell about 10 percent, and the earning of high school dropouts fell by 25 percent.

Large tech companies are another type of knowledge owners, and the rise of their value reflects the sea change from things to thoughts. The shift has created unimaginable wealth for knowledge owners. In 2017, five of the five biggest companies in the world were knowledge driven—Apple, Alphabet (Google's parent), Microsoft, Amazon, and Facebook. In 2011, Apple was the only one in the top five and in 2006, only Microsoft was a top-fiver; the number one in 2006 and 2011 was Exxon Mobil (Table 3.1).[9]

8. David Autor, "Skills, Education, and the Rise of Earnings Inequality among the 'Other 99 Percent,'" *Science* 344, no. 6186 (2014): 843–851.

9. Antoine Gourévitch, Lars Fæste, Elias Baltassis and Julien Marx, "Data-Driven Transformation: Acclerate at Scale Now," Boston Consulting Group blog, May 23, 2017.

Table 3.1 Top-Ten Largest Companies by Market Capitalization: Recent Dominance of Knowledge-Driven Firms

Stock Market Rank	2017	2011	2006
1	*Apple*	Exxon Mobil	Exxon Mobil
2	*Alphabet (Google)*	*Apple*	General Electric
3	*Microsoft*	PetroChina	*Microsoft*
4	*Amazon*	Royal Dutch Shell	Citigroup
5	*Facebook*	ICBC	Gazprom
6	Berkshire Hathaway	*Microsoft*	ICBC
7	Exxon Mobil	*IBM*	Toyota
8	Johnson & Johnson	Chevron	Bank of America
9	JPMorgan Chase	Walmart	Royal Dutch Shell
10	*Alibaba Group*	*China Mobile*	BP

* Data-driven companies

SOURCE: Author's elaboration of data published in *BCG Perspectives*, 2017.

An additional source of fuel for the upheaval came from a shock rise in income inequality. The transformation of advanced economies from industrial to post-industrial societies has not been gentle on the "forgotten men and women."

Economic Inequality

In the United States, the pattern is very clear and very pronounced. The well-off did well, the poor did poorly, and the average did awfully. The average US man working full-time got $53,000 in 1973, but only $50,000 in 2014 in inflation-adjusted terms.[10] The average American family is sliding backward in terms of earning power—and has been since the early 1970s. Only half the population has seen incomes rise over the past three decades. The incomes of the other half have fallen. And even among the winners,

10. Here "average" means "median," i.e., the earner that is exactly halfway up the income ladder.

the winnings have been astoundingly concentrated in the pockets of the very richest. The bottom 90 percent's share of the American economic cake, which had been about two-thirds during the thirty glorious years, rocketed down to a half by the 2000s.

In Britain, the share of national income going to the top 1 percent income bracket more than doubled from 6 percent to 14 percent. Curiously, this is not what happened in the rest of Europe or in Japan. In these nations, inequality tended to fall from the 1970s to the 1980s, before rising. They are now back at their 1970s starting point and seem stable.

The causes of these varied changes in income equality are many and complex. While this has been a topic in seminar rooms for many years, it burst into the open with the 99 Percent movement; the Occupy Wall Street movement; and Thomas Piketty's transformative 2013 book, *Capital in the Twenty-First Century*. The explanations range from government deregulation and the rise of monopoly capitalism to the decline of labor unions and skill-biased technology progress.

Technology surely played a role. Many elements of the ICT impulse tended to boost income and wealth inequality. The skill twist, for example, meant that the wages for higher income earners were favored over those of the working class. People with higher levels of education started with higher incomes and saw them get higher swiftly. This dynamo worked in reverse for high-school-only people. Their incomes started lower and went even lower. The shift in value creation and capture from capital to knowledge created a new class of super-rich.

Since there are a lot of people in the low education category, the gigantic gap between productive growth and wage growth has swallowed hundreds of millions in Europe and, especially, America. There, the combination of income stagnation, the destruction of good industrial jobs, and long-running decimation of communities that used to thrive around manufacturing hubs has yielded some very bad non-economic problems.

The massive economic transformation that came with the ICT-led automation and globalization produced backlashes in America and Europe. The 2016 backlash is nowhere near as big as the great backlashes of the early 1900s. It is more like the small backlashes of the early 1800s—the

Luddites and Corn Laws—but we don't yet know where it is heading. The surprise election of the populist outsider Donald Trump as president was the largest backlash so far.

NEW UPHEAVAL PRODUCES A NEW BACKLASH

Donald Trump got Jeff Fox's backlash vote, but not for the reason you might expect given the economic hardships he faces. Fifty-eight years old, he is a cancer survivor with a massive healthcare debt, living on disability and social security payments. While his father was an accountant in Bethlehem Steel—the region's economic powerhouse until its 2001 bankruptcy—Fox was a furniture salesman before his early retirement. His daughter worked at Walmart. "We have voted with our principles and our conscience for all these years, and where has it gotten us?" questioned Fox.

Other voters backed Trump just to shake things up. Duane Miller, owner of a paint and wallpaper store and former Democratic mayor of the small town, Bangor, Pennsylvania, said: "It's the disillusionment of the common man with government, because government has done nothing to help the average working man." He continued, "The political climate for the average American, from my point of view here in the little town of Bangor, is one of disbelief. The American people don't believe anything anymore. And that's where the apathy is overwhelming."[11]

At one level, the 2016 election of an autocratic outsider promising to restore strength and stability is easy to understand.

Interpreting the US Backlash

As in the 1920s and 1930s, many Americans felt left behind in 2016. Rapidly advancing automation in manufacturing combined with the offshoring of

---

11. Tom McCarthy, "Trump Voters See His Flaws but Stand by President Who 'Shakes Things Up,'" *The Guardian*, December 24, 2017.

industrial and back-office jobs to create a systematic and very persistent threat to workers in the middle of the skill range. Many of the displaced workers have found work but in much lower paying, more precarious positions.

Deindustrialization has destroyed communities, and people are reacting as members of threatened communities, not just individuals whose jobs are at risk. People are finding that they cannot afford to a buy a house like the one they grew up in. Many millennials find themselves weighed down by student debt, right when the new economy has meant that a university education is no longer a sure ticket to a middle-class lifestyle. And things are evolving so much faster.

Since the changes are more sudden, more individual, more unpredictable, and more uncontrollable than before, economic fragility is back. Once again, job loss can have dire consequences; unemployed Americans risk losing their homes and healthcare. After having given Republicans and Democrats eight years each to fix the problem, minds were open to more unconventional solutions. Trump's narrow victory, however, has many complicated facets.

While decades of declining fortunes primed people like Fox to go for an outsider like Trump, his was not a vote for European-style social welfare. "It would be nice for me to say, I got $40,000 of medical bills, so it'd be nice if someone paid them for me," Fox explained, but continued, "It's not the responsibility of the government to pay the bills."

Trump's victory is a delicate thing to understand. He is no FDR. Roosevelt had a plan to help people and proven track of having done so (as governor of New York State). The policy FDR implemented in New York was a model for the New Deal.

Trump, by contrast, didn't have a plan to uplift the downtrodden, and certainly no track record. He had slogans and a bully's attitude. His program was ill specified, and incoherent on many levels. But his rhetoric was combative and patriotic. Moreover, his win rested on a razor's edge.

He lost the popular vote by 2.9 million votes (2 percentage points). His electoral college vote came down to seventy-seven thousand ballots in three states (all hard hit by the Services Transformation). If twenty-three thousand Pennsylvanians, twelve thousand Wisconsinites, and six thousand Michiganders had switched their votes, Hillary Clinton would have been elected president.[12]

This was not an FDR-like upwelling of discontent. Less than 60 percent of eligible voters even bothered to fill out a ballot. Economic and social calamity had been swirling around the country for years. Many low-skill white men outside of large urban areas have been left behind by the post-industrial society, and this group voted heavily for Trump. People who said their family's financial situation was worse in 2016 than 2012 voted heavily for Trump (78 percent), while only 39 percent of those who reported things being about the same did.[13] Those who thought the nation's economy was in a poor state voted for Trump, as did 65 percent of those who thought trade takes jobs away. Personal income, however, was not a reliable predictor of Trump voting. More than half of people who were forty-five or older voted for him, while less than half of those under forty-five did. More than half of those with less than a college education voted for him; less than half of those with a college education did.

But surely it was more than a matter of economics. In fact, many social scientists have a different take on the Trump triumph.

Political scientist Karen Stenner argues that Trump is riding a wave of autocrat-seeking voters—voters who want strength and order to counter the drift and hopelessness they and their parents have experienced since the 1970s. They want "to make America great again." Stenner sorts Trump voters into three bins: "economic conservatives" who embrace private entrepreneurship, large corporations, free markets and free trade; "status quo lovers" who just don't like change of any kind; and "authoritarians" who only get riled when they think their communities are menaced, and the current leadership is unwilling or unable to fix the situation.[14]

12. *Business Insider*, 2016 election exit polls,*uk.businessinsider.com*.

13. "Election 2016: Exit Polls," *New York Times*, August 11, 2016.

14. Antoine Gourévitch, Lars Fæste, Elias Baltassis and Julien Marx, "Data-Driven Transformation: Accelerate at Scale Now," Boston Consulting Group blog, May 23, 2017.

John Jost, an New York University professor of psychology, notes that Trump's personal style—while abhorrent to many—is powerfully attractive to the authority-seeking voters, including many—like Duane Miller—who voted Democratic previously. When Trump bullies political opponents and the press, he is tapping into a deep well of resentment of the establishment that let America go so wrong for so long. His swagger, refusal to play by the rules, refusal to apologize, and absolute self-confidence are balm to this sort of voter.[15]

Brexit

The June 2016 British vote to leave the European Union (EU) was, if anything, even more shocking than Trump's victory. For one thing, it was the first concrete sign that a backlash was under way in 2016. And it was unexpected.

Few people "in the know" expected the sensible, cautious Brits to take such an incredible leap into the unknown. EU rules and practices were—after four decades of knitting—woven throughout Britain's entire economic and regulatory fabric.

The real problem with the referendum was that it unified voters' discontent without clarifying their intent. The referendum asked voters whether they wanted the country to embark on a grand voyage without specifying the destination. The entire text of the question was: "Should the United Kingdom remain a member of the European Union or leave the European Union?" The possible answers were just: "Remain a member of the European Union," or "Leave the European Union."

While the implications of "remain" were absolutely clear—it was what people had known for over forty years—the meaning of "leave" was absolutely unclear. The "leave" campaign could not agree on what sort of economic, political, and security relationship the United Kingdom should

15. See interview with Jesse Graham, a professor of psychology at the University of Southern California in Edsall, "Purity, Disgust and Donald Trump," *New York Times*, June 1, 2016.

have outside the EU. Different "leave" campaigners promised different things.

The ruling Tory Party was so badly divided on Brexit that the critical issue of Britain's post-Brexit trade relationship with the EU didn't come up for a Cabinet discussion until eighteen months after the vote. And this despite the fact that the United Kingdom does more than half its trade with the EU. When this book went to press, Tory Party members firmly agreed that they should exit the EU, but they still had not agreed on where they were going to exit to. Intra-party splits prevented the Tories from agreeing among themselves on what sort of long-term trade relationship they wanted with the EU. This makes the whole backlash look a lot more like a cry of anguish than a clear call for the fundamental way the UK economy is run.

The nature of the Brexit backlash was quite different from the US election of Trump—it was not at all about electing a strong, autocratic leader in time of peril. While there was a good deal of nationalistic drum-beating during the campaign, and subtle racist undertones, none of the pro-Brexit campaigners could be considered strong, charismatic leaders. And in any case, once the leave camp won, all its leaders walked off or were pushed off the stage.

The thankless task of implementing the will of the people was left to an oddly awkward politician who actually voted against Brexit—Theresa May.

While it is very hard to know exactly what voters wanted, it is quite easy to understand the discontent that drove their votes.[16] There was certainly an element of protest vote, or cry of anguish, to the outcome. An exit poll showed that 70 percent of voters thought the remain-in-the-EU side would win—including 54 percent of those who voted to leave. Voting patterns quite neatly mapped out the regions and demographic groups most harmed by the Services Transformation. People who had faced prolonged hardship wanted to leave; those looking to the future wanted to remain.

16. Lord Ashcroft, "How the United Kingdom Voted on Thursday. . . and Why," *lordashcroftpolls. com*. June 24, 2016.

The same exit poll showed that leave voters were older, less educated, and more likely to be living outside major urban areas than remain voters. Almost three-fourths of eighteen to twenty year olds voted to remain, sixty percent of twenty-five to thirty-four year olds wanted to stay, but a majority of those aged over forty-five voted to leave. Fully 60 percent of those beyond retirement age wanted out. A majority of voters with jobs voted to remain, but a dominant majority of the unemployed voted to leave. A large majority of people with high school degrees or less voted to leave.

Importantly, it was not a vote defined by party affiliation. While 40 percent of leave voters associated with the Conservative Party, half as many identified with the Labor Party. Indeed, both mainstream parties were torn internally over the decision. Only the far-right, pro-leave UK Independence Party was cohesive, and it disintegrated as a political force once the referendum was over.

While Brexit and Trump's unexpected victory primed 2016 to be a turning point, other European electorates didn't comply.

The European Continentals That Didn't Lash Back

In non-UK Europe, right-wing, populist parties have long existed alongside the mainstream left–right political divide. They are fringe parties and consider themselves as such, with vote shares hovering between 5 and 20 percent. This changed in the 2010s. The 2014 elections for the European Parliament saw a rise in vote-shares for anti-EU parties in most EU nations, including the Big 4: France, Italy, Germany, and Britain. Overall, these far-right populist parties saw their share rise from under 20 percent to over 30 percent between the 2009 and 2014 elections.

At the national level, a worryingly far-right candidate, Marine Le Pen, looked set to win the French presidency, and poll-numbers of populists in several other nations surged. In the end, the French strongly rejected the French version of Trump. Dutch populist Geert Wilders's party, the Party for Freedom, did well but didn't win. The antimigrant, populist upstart party, Alternative for Germany, did well enough to get 13 percent of parliamentary seats, but it didn't enter into power. In Austria, the far-right Freedom Party entered into a power-sharing arrangement in 2017 and is thus part of the government. Yet, this was not a populist upheaval. Austrian soundly rejected the far right in the December 2016 presidential election. Instead, they went for a former Green Party leader, Alexander Van der Bellen, who styled himself as "open-minded, liberal-minded and above all a pro-European."[17]

The key to understanding what happened in Europe is to distinguish sharply between antiglobalization and antimigration sentiments.

The 2016 and 2017 surges in far-right voting were largely unconnected to the lingering middle-class malaise that was so important in the US and UK. Much of it was directly tied to the European refugee crisis that started in 2015 and saw the arrival of something like 1.5 million immigrants from Syria and North Africa. And trust was a big driver.

A recent study by leading economists showed that "lack of trust in national and European political institutions" was the common thread through European populism. They found that it was the old and the less-educated who were driving the trend. This suggests that some of the things that drove US and UK backlashes were also important in Europe, but things are nowhere near as extreme. As the 2017 report, *Europe's Trust Deficit: Causes and Remedies*, puts it, the research results "do not suggest that there is a real and present danger of the EU disintegrating. The UK is an outlier. The crisis has left a toll, but the effects of negative macroeconomic shocks on attitudes towards the EU are not very large. And with economic conditions now improving, attitudes and electoral outcomes ought to turn more favorable to the EU, assuming that history is a guide."[18]

When this book went to press in mid 2018, this judgment seems to be holding up well. It suggests that 2016 was, like 1848, a historical turning point where history failed to turn.

---

17. Philip Oltermann, "Austria Rejects Far-Right Candidate Norbert Hofer in Presidential Election," *The Guardian*, December 4, 2016.

18. See Christian Dustmann, Barry Eichengreen, Sebastian Otten, André Sapir, Guido Tabellini, and Gylfi Zoega, "Europe's Trust Deficit: Causes and Remedies," *VoxEU.org*, August 23, 2017.

Some of the most revealing pieces of the 2016 backlash puzzle come from what happened in Japan. Or more precisely, from what didn't happen in Japan.

Japan's Missing Backlash

The Services Transformation hit Japan as hard as any nation on earth. Maybe even harder since its economy was so reliant on manufacturing. Japan's thirty glorious years, which were more glorious than Europe's and America's, were followed by the "Lost Decades." Indeed, Japan has suffered one of the longest economic crises in history. Its economy actually shrank by a fifth between 1995 and 2007. Part of this came from falling prices and a declining workforce, but real wages did fall by 5 percent.

Despite the economic hard times, the Japanese people are pro-globalization. A recent Pew Research poll found that 58 percent of Japanese agreed that involvement in the global economy "is a good thing because it provides Japan with new markets and opportunities for growth." Only 32 percent said that "it is a bad thing because it lowers wages and costs jobs."[19]

The key difference between the United States and Japan, in my view, is the cohesiveness of the society. The Japanese understand that pains and gains come as a package, but they expect that both the pains and the gains will be shared. They believe their leaders are working in their best interest.

A telling example is the populist backlash that backfired.

In the crazy days of late 2016 and early 2017, when politics in advanced economies seemed to have been turned on its head in the US and Europe, a populist politician in Japan stepped up with hopes of upsetting the establishment. The sitting prime minister Shinzo Abe announced surprise elections and one of his former allies, Yuriko Koike, announced a surprise of her own. The highly popular sitting governor of Tokyo quit the ruling

19. Bruce Stokes, "Japanese Back Global Engagement Despite Concern about Domestic Economy," Pew Research Center, October 31, 2016,

party, set up the "Party of Hope," and declared her intention to unseat the incumbent prime minister.

Her campaign talk was straight out of the populist playbook, which claims that, as I phrase it: "The people are pure, the elite are corrupt, so vote for me so I can fill-in the-blank." The fill-in the-blank part is not very important. In Koike's case, she described herself as conservative populist, claiming: "If at this time we don't reset Japan, we won't be able to sufficiently protect our international competitiveness and national security."[20]

The new party blew up the old alternative party, the Democratic Party, and attracted several high-profile conservative politicians. The media drew strong parallels with Brexit, Trump, and European populists like Marine Le Pen. It looked like the backlash that started in 2016 would continue into 2017 in Japan. In the end, little came of this challenge.

Koike won only half the votes Abe did. Abe's traditional party, the Liberal Democratic Party, not only won the election but won more than two-thirds of the parliamentary seats, which gave Abe the supermajority he needs to reform the constitution. The attempted populism, in other words, had the effect of handing even more power the the establishment. Koike went back to being governor of the Tokyo region.

THE MISSING RESOLUTION AND THE NEXT TRANSFORMATION

New Deal capitalism ushered in economic contentment and broad-based prosperity. Incomes soared on the back of technological progress and expanding trade—especially for the middle class. FDR's "forgotten" men and women were forgotten no longer. They saw life-changing increases in living standards, financial security, and economic prospects.

This happy position started to slip in the 1970s as the nature of technological progress changed. Manufacturing employment in the US peaked

---

20. Elaine Lies, "Tokyo Governor Launches New Party, Won't Run for Election Herself," *Reuters. com*, September 27, 2017.

in 1979. Due to automation, it has trended downward ever since. And then came the new globalization around 1990. This tipped rich nations' share of world manufacturing into a steep decline—one that continues today.

The massive economic transformation that came with ICT-led automation and globalization—above all the deindustrialization and slow growth—produced a backlash and unfocused calls for shelter from the shocks. The backlash is nowhere near as big as the great backlashes of the early 1900s, but we don't yet know where all this anger is heading. A key point to keep in mind is that the 2016 backlash has not produced a resolution. Nothing substantial has been done to redress the underlying misery, insecurity, and generalized sense of fragility that permeates many layers of society. This is especially true in the US, and, to a lesser extent, the UK, but elements of the malaise exists in all the advanced nations.

A new technological impulse—digital technology—has hit the world and launched an economic transformation. This is really something new due to the volcanic pace of the technological progress. Things that seemed implausible last year—like instant, free translation—are ubiquitous today. This is not evolution with the fast-forward button pushed. It is really something different. It is a technological revolution of sorts—a fact that many have missed.

PART II

The Globotics Transformation

4
—

The Digitech Impulse
Driving Globotics

Mike Duke was in denial about the explosive pace of digitech, but no longer. "I wish we had moved faster," said the former CEO of Walmart. "We've proven ourselves to be successful in many areas, and I simply wonder why we didn't move more quickly." Mickey Drexler, CEO of clothing retailer J.Crew, expressed a similar sentiment a month before "former" was added to his title: "I've never seen the speed of change as it is today. If I could go back 10 years, I might have done some things earlier."[1]

The speed of change is clearly hard to comprehend. Many people are either unaware of how fast the changes are coming or are living in denial. The US Secretary of the Treasury, Steve Mnuchin, is in the unaware camp.

Asked in March 2017 whether AI would replace workers, Mnuchin responded: "I think that is so far in the future. In terms of artificial intelligence taking over American jobs, I think we're like so far away from that, that uh [it's] not even on my radar screen. Far enough that it's 50 or 100 more years." This quote is illuminating since Mnuchin is not some hapless soul who watches too many segments about World War II on the

1. Khadeeja Safdar, "J.Crew's Mickey Drexler Confesses: I Underestimated How Tech Would Upend Retail," *Wall Street Journal*, May 24, 2017.

History Channel. His ability to see the future has paid off handsomely in the past.

In 2009, in the depth of the global crisis, Mnuchin bought a failed mortgage lender and pocketed a billion dollars in profit when he resold it in 2015. This guy is so rich that in the financial disclosures he had to fill out to become treasury secretary, he left off over a hundred million dollars in wealth by accident. When pressed at his congressional hearing, he explained: "I think as you all can appreciate, filling out these government forms is quite complicated."[2]

There are good, deep-seated reasons why people as sophisticated as Duke, Drexler, and Mnuchin have trouble understanding the inhuman pace of digitech. Explosive growth is something our walking-distance brains have trouble comprehending. Think of it as the unintended consequence of an evolutionary hangover.

BRAIN BUG VERSUS EXPONENTIAL GROWTH

Our brains are the key bit of equipment when it comes to thinking about the future of digital technology, but our brains evolved to do something quite different. All animal brains, including ours, evolved to track motion. Things that move have brainpower; things that don't, don't. There is even an animal—the sea squirt—that has a brain when it is in its mobile life phase, but loses it once it is permanently attached to something.

This matters since the evolution took place in a very different world—a walking-distance world. We thus have a strong tendency to assume that things that changed between yesterday and today will change between today and tomorrow at more or less the same pace. We are primed by evolution to make straight-line extrapolations when thinking about the future.

2. Alan Rappeport, "Issues of Riches Trip Up Steven Mnuchin and Other Nominees," January 19, 2017, New York Times. For his quote on AI, see Shannon Vavra, "Mnuchin: Losing Human Jobs to AI 'Not Even on Our Radar Screen,'" www.axios.com, March 24, 2017.

Many of us think of ourselves as thoroughly modern, but in reality, it wasn't that long ago that bows and arrows were hi-tech weapons. People started living in cities only about six millenniums ago. Six thousand years sounds like a long time in a world where watching the first five seconds of an ad on YouTube seems like an unreasonable imposition. But it is actually not that long—not on the evolutionary timescale. Think of it this way.

Imagine you could gather your ancestors for a reunion—your mother, your grandmother, your grandmother's mother, and so on, back to the days when the first humans lived in cities. How much wine would you have to order for this grand reunion? The answer is surprisingly little. You could fit the whole party into a big movie theater with room to spare. There would only be three hundred of you. If they were all polite drinkers, which means a quarter bottle each, you'd have to lay in only a dozen crates, seventy-five bottles in all. The point is plain.

In evolutionary terms, three hundred generations is not much more than the five-second ads on YouTube. This is why our brain is not really fit to deal with the globotics upheaval. Our brains evolved to understand straight-line growth in a world where really fast meant a spear in flight. But digital technology doesn't fly that way.

How Digitech Ambushes Our Walking-Distance Minds

Digital technology advanced by small increments at first since it started from zero. For years, the progress was almost imperceptible, but then the increments got immense—a pattern we can illustrate with an example from banking.

If a bank account paid the extremely high interest rate of 58 percent per year, your money would double every 18 months and that means a penny deposited today would be worth a dollar in ten years. That's a hundredfold increase, but a dollar from a penny is hardly earth-shaking. That's growth in the "imperceptible progress" phase.

Things would be more exciting in the second and third decades, but the fourth decade is when the increments would start to impress; you would see 10 thousand dollars turn into a million dollars in the fifth decade. After that, the increments get implausibly immense. Your million becomes 100 million in the sixth decade, and 10 billion in the seventh. That's the "explosive progress" phase.

That sort of growth seems strange: a penny into ten billion dollars with the progress being way below the radar screen for thirty years. That just doesn't seem normal, and it's not if you are straight-lining the future. But it is exactly how exponential growth works. It is exactly how digitech is advancing. And it is this imperceptible-for-decades-then-explosive feature that makes it so hard to think intuitively about digitech's exponential growth.

Take computer processing speeds, for example: they are doubling every 18 months or so. The iPhone 6s, which came out in 2015, processes information about 120 million times faster than the mainframe computer that guided Apollo 11 to the moon in 1969. That is amazing. But it gets more amazing. The iPhone X, which came out in 2017, is about three times faster than the iPhone 6s. That means the increment in processing speed between 2015 and 2017 was 240 million times the speed of the Apollo 11 computer.

Think about that. The increment in power in the two years after 2015 was twice as large as all the progress between 1969 and 2015. Twice as much progress in two years as there was in the 46 previous years. That just does not seem normal to our walking-distance brains. This imperceptible-for-decades-then-explosive feature is why many are either unware of how fast the changes are coming or living in denial.

We can draw a picture of this mismatch between our natural tendency to straight-line the future and the actual shape of the exponential growth. I call it the "holy cow" diagram.[3]

3. I was inspired in drawing this by a blog post by Ro Gupta, "Why We Overestimate the Short Term and Underestimate the Long Term in One Graph", *www.rocrastination.com.*

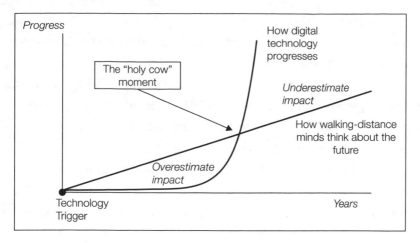

Figure 4.1 The Holy-Cow Diagram.
SOURCE: Author's drawing.

The "Holy Cow" Diagram

Our intrinsic tendency to straight-line the future is illustrated with the straight line that rises steadily from left to right (Figure 4.1). The actual way that digital technology progresses is shown as the hockey-stick-shaped curve. During the imperceptible-progress phase, it is bumping along the bottom. When it hits the explosive-progress phase, it rockets upward as shown.

When the explosive growth of digital progress crosses the human projection of progress, we get what I think of as the "holy cow" moment. This is when digitech is "disruptive". People knew it was coming—they just didn't expect it to come so fast. They just can't comprehend why things are changing so fast now when they weren't changing that fast in the past.

The progress during the explosive growth phase just doesn't seem feasible or reasonable given past experience. And in a walking-distance world, it isn't reasonable. In an exponential growth world, by contrast, it is inevitable—as the ex-CEOs Duke and Drexler found out the hard way.

There is another way to highlight the disconnect between intuition and reality when it comes to digital technology—it's called Amara's law. The futurist Roy Amara said we tend to overestimate the effect of a technology in the short run (before what I call the "holy cow" moment) and we underestimate the effect in the long run. This rather systemic miscalculation is not a new thing.

Pierre Nateme, CEO of Accenture, wrote in 2016: "Digital is the main reason just over half of the companies on the Fortune 500 have disappeared since the year 2000." And digitech it is not just affecting companies—it is transforming the world of work.

When did the new impulse begin? Dating a revolution like this one is impossible since revolutions are processes, not events. That said, 2016 or 2017 are good guesses. Let's just say 2017 since that was "The Year of AI" according to the Forbes Technology Council, and Fortune magazine.

But what is this digital technology?

FOUR DIGITECH LAWS

Digital technology is really quite unusual. The way it progresses is so remarkable that it has special names. Moore's law, which is one of these special names, states that computer processing speeds grow exponentially, doubling every 18 months or so. There are three other "Laws" that explain the unusual nature of digital technology. The one about the growth in data transmission is called Gilder's law, the one about the growth in the usefulness of digital networks is called Metcalf's law, and the one that explains the insane pace of innovation is called Varian's law. The people behind the laws are as interesting as the laws themselves.

Moore's Law

Gordon Moore's career is, in a strange way, an analogy for how his law works. He started slow but went on to do amazing things. An indifferent student in high school, he spent two years in the unglamorous San Jose State University before transferring to the big leagues at University of California—Berkeley and becoming the first member of his family to graduate from university. He started his work on semiconductors under the guidance of the inventor of the transistor, William Shockley. Things did not go well at Shockley Semiconductor.

Shockley was a rare character. A difficult man to work for in the best of circumstances, his behavior became increasingly erratic and autocratic after he won the 1956 Nobel Prize in Physics. Soon after, Moore and seven other young researchers left to form their own company. With seed capital of $500 from each of the eight—and backing from Fairchild Camera and Instrument—Fairchild Semiconductor Corporation was born in 1957. Moore was the R&D director and published his famous law in 1965. After a decade at Fairchild, Moore left to start up Intel Corporation in 1968. That made him a billionaire, and earned him the Presidential Medal of Honor.

Moore retired in 1997, but his law kept rolling. The number of transistors per square inch has doubled approximately every eighteen months since Richard Nixon was president. One reason was that it soon stopped being something that chronicled progress and became something that drove it.

One key point about Moore's law is that it is not a law like the law of gravity. It is not even a rule of thumb. Rather, think of it as a rallying cry or the official anthem of the electronics and software industries. It orchestrated progress for five decades.

Orchestration in the IT world is needed since the companies that make the chips don't design the software and computers that use the processing power. It's a bit like the relationship between jet engine makers, like Pratt & Whitney and Rolls Royce, and aircraft makers like Boeing and Airbus. The jet makers spend years and millions developing jet engines for planes that don't yet exist, while the plane makers spend years and millions developing planes that won't fly without engines that do not yet exist. This coordination is not difficult since there are so few firms involved, but the IT industry is global and ever-changing.

IT companies invest millions of dollars for years to develop breakthrough software and telecommunication services that can only work on computer chips that don't yet exist. Likewise, chipmakers invest hundreds of millions for years to designing better chips in anticipation of the frothy demand that flows from the breakthrough software and telecommunication services that come online every year. To put it differently, Moore's law is a self-fulfilling prophecy, or maybe even a Ponzi scheme.

For decades, the home of Moore's law, and the coordinating mechanism for chip makers and users, was the *International Technology Roadmap for Semiconductors*. The 2015 report, which was the last, predicted Moore's law would continue apace until at least 2020. No one thinks this will be easy or automatic.

Recent research shows that it now takes seventeen times more research hours to double processing speeds than it did in 1971. This means that the sums at play are enormous. The specialty chipmaker, Nvidia, for example, spent over two billion dollars developing a new chip that speeds up machine learning. That is a lot of money. It would, for instance, pay for half of a US Navy Nimitz-class nuclear aircraft carrier. And all this for a chip that makes machine learning about twelve times faster.

The reason such sums make sense is that the demand for faster chips is growing equally fast. That, ultimately, is why Moore's Law continues to bind—people are still making money selling faster chips.

Gilder's Law

As with Gordon Moore, there is a strange parallel between George Gilder the man and the law he named after himself. In 1989, Glider predicted that data transmission rates would grow three times faster than computer power. This prediction went through a massive hype cycle—a bit like Gilder himself. The two stories are surprisingly intertwined.

The technology breakthrough that triggered the hype cycle was the commercial viability of fiber optic cables. These promised vastly faster transmission rates. The innovation was oversold at first, largely by Gilder himself. This fostered overinflated expectations that became part of the "dot com" bubble of the late 1990s. Data transmission speeds did grow much faster than processing speeds for a few years, but then slowed to about the same pace as Moore's law.

Gilder got the investment side of the technology terribly wrong—enough so that it bankrupted him personally when most high-tech stocks crashed in 2001. But his predictions of explosive growth in transmission have come true—but not quite as he predicted, as Figure 4.2 shows. Until the mid-1990s, the internet in the US was government controlled. Despite an explicit policy of discouraging commercial activities, it grew at over a hundred percent annually—which is about three times faster than processing power was growing. When the internet was privatized in 1995, it exploded—growing at almost a thousand percent per year in 1995 and 1996. After that, the growth rate gently declined and is now in the solid double-digit range, say 20 to 30 percent per year. The result is that today an absolutely insane amount of information is transmitted daily.

In a single typical minute in 2017, a half million Tweets were sent, over four million YouTube videos were watched, 47 million Instagram posts and 4 million Facebook likes went up, and 15 million text messages were sent. To talk about the total volume of data transmitted in 2016, you need words you probably have never heard before. Cisco estimates that global internet traffic was 1.2 zettabytes in 2016. That is a very large number.

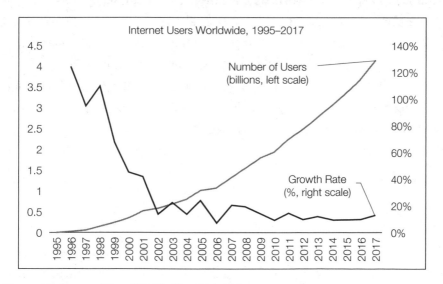

Figure 4.2 Internet Users Worldwide, 1995–2017.
SOURCE: Author's elaboration using data published on World Internet Stats.com, https://www.internetworldstats.com/emarketing.htm.

It takes eight bytes to store the letter 'a', or indeed any other character. Storing all the catalogued books in the world in all languages (plus a backup copy) would fill about 480 million million bytes. That's 480 followed by 12 zeros and would fit neatly on to about 20,000 DVDs. Stacking those would produce a pile that's about 24 meters high. A zettabyte is a trillion times more than that. Storing 2016's internet traffic on DVDs would require a stack that is 24 billion kilometers high. The sun is only 150 million kilometers away, so the stack would reach from the earth to the sun and back 80 times.

And the numbers are climbing rapidly. Cisco estimates that the amount of information crossing the internet will double every couple of years up to 2021. In addition to individual connections getting faster, the number of connections has risen rapidly worldwide. In its early days, the number of internet users exploded—rising at triple-digit growth rates. That calmed down to the ten to twenty percent range from about 2000, where it has stayed ever since. Now there are over 4 billion users. The coverage is close to complete in North America and Europe. In Asia and Africa, however, there is plenty of room for growth as less than half the world's population is online. For the world as a whole, internet connectivity is at about 55 percent. At the current growth rate of about 10 percent per year, over a billion more people will be online by the time the US election rolls around in 2020. By the 2024 election, almost every human will be online.

The combination of fast data processing and fast transmission has produced some absolutely enormous digital networks, like Facebook with its two billion users. There is a very good reason for this—it's called Metcalf's law.

Metcalf's Law

Robert Metcalf—the third and least colorful of the digital lawmakers—observed that being connected to a network gets more valuable as the network grows, even as the cost of joining falls. This not only helps explain why digital networks grow so fast, it also explains the winner-take-all outcomes we see with online competition among networks. The law is really just common sense.

It is pretty obvious that networks are more useful, and useful more often, when they link-up more people, more computers, and more information. But the simple trend—more links means more useful—is not where the insight lies. Metcalfe's law states the value of a network grows faster than the number of people connected to it. And not just a little bit faster, it grows twice as fast.

When the number of network users is, say, 100,000, the number of possible new connections created by adding one more user is 100,000. When there are 200,000 users, adding one more creates 200,000 new connections. In other words, the incremental number of new connections does not rise in a straight line. The size of each increment grows with each new increment, so growth feeds on growth and soon can become transformative.

The outcome is sometimes called "tipping-point economics". When the size of a thing gets past its tipping point, it can snowball into something very big, very fast. WhatsApp is a good example. People started joining in droves since people started sending lots of messages and the larger audience, in turn, spurred more people to send more messages. In the 16 months leading up to July 2017, an extra half billion people started using WhatsApp. The snowball effect also has a social element to it. People often don't do something because other people don't do it. But when others start doing it, many join in.

The essential point is that networks get more valuable much faster than they get big. This has a few important implications for the age of globotics.

The first is that it helps explain why the economy in cyberspace seems to act differently than the economy in real space. It helps explain why companies like Facebook, WhatsApp and Twitter can get so valuable so fast. Facebook, to take the classic example, was launched in 2004, but was only opened to the public in September 2006 (initially it was only for university students). Five years later it had 600 million users. In 2012, it earned a billion dollars in profit. Today, it has over 2 billion users and earns over $10 billion annually.

The second point is that Metcalf's law helps explain the tendency of the virtual economy to act as a winner-take-all contest. In the 2000s, Facebook had a few competitors, like MySpace, but everyone wanted to be on Facebook since everyone else was on Facebook; that was where you could find your friends. Likewise, I can remember when Google was the new search engine challenging incumbents like Yahoo. Victory was not all assured but once Google started winning, it gained users that made it win faster. Lycos, Altavista, Ask.com and the like all went by the wayside. Even a search engine "born big," like Microsoft's Bing, has trouble challenging the leader's primacy due to Metcalf's law.

The power of networks and the eruptive pace of raw computing and transmission power are not the only thing driving the inhumanly fast pace of digitech. There is something very different about innovation in the digital world compared to the industrial world. The cluster of new technologies that arose in the late 1900s during the Second Industrial Revolution took decades to generate useful products and new processes. The invention process was slow since inventing involved physical things.

The nature of digital innovation is quite different. It is radically faster because the nature of the underlying components is so different. There is even a name for this new type of innovation—digital, combinatoric innovation. That's what Hal Varian, the Chief Economist of Google, calls it, but I think of it as Varian's law. In some ways, Varian as a person is quite different than his law.

Varian's Law

Hal Varian is a tall, laid-back man who looks a decade younger than his 70 years. His law is all about chaotic innovation, but there is nothing chaotic about him—unless you count his mischievous sense of humor. When I spoke with him at the ECB Forum on Central Banking in Sintra, Portugal, in the summer of 2017, he seemed to be gently poking fun at the central bankers assembled. His dried-orange-peel colored tie (surely made of California's finest polyester) was emblazoned with $, £, €, and ¥

symbols imposed on line charts representing stock markets booms and busts. Maybe this was a Silicon Valley salute to the ECB's formal dress code, or maybe he was just planning to use it as an ice-breaker with Mario Draghi and Ben Bernanke.

Varian's law explains why things are changing so fast these days in the digital world. "Every now and then a technology, or set of technologies, comes along that offers a rich set of components that can be combined and recombined to create new products," explained Varian. "The arrival of these components then sets off a technology boom as innovators work through the possibilities."

The big difference between today and the 19th and 20th century innovation booms is the nature of the products and the components. Today the components are things like open-source software, protocols, and Application Programming Interfaces (APIs). Strange as it may seem, these components are free to copy.

Even in the competitive world of machine learning, the leaders are publishing their key research findings in academic, open-access journals. Large training datasets are routinely posted for free downloading. Companies like Google have made their most powerful computers free to use for some online users. IBM has made its cutting-edge quantum computer available for free in order to create a community of experts who know how to get quantum computers to do useful things.

What might seem strange about this widespread practice is that the digital products made of these free components are often insanely valuable.

Varian's law is thus: digital components are free while digital products are highly valuable. Innovation explodes as people try to get rich by working through the nearly infinite combinations of components in search of valuable digital products.

In their breakthrough book, *The Second Machine Age*, Erik Brynjolfsson and Andy McAfee point out the implications. A big difference between digital technology and traditional technology is that new products and components can be reproduced costlessly, instantly, and perfectly. Imagine how much faster the Industrial Revolution would have spread if Newcomen's steam engine could have been reproduced costlessly, instantly, and perfectly.

Self-driving cars are an example of Varian's law. They are one of the sure-fire, high-tech wonders of the future. Yet they use no breakthrough technology. They are a recombination of existing technologies like GPS, Wi-Fi, advanced sensors, anti-lock brakes, automatic transmission, traction and stability control, adaptive cruise control, lane control, and mapping software—all integrated by tons of processing power, and an AI-powered white-collared robot. Yet, despite being a mash-up of off-the-shelf tech, self-driving cars will create a $7 trillion market. This is not an isolated example. Many of today's most innovative products, apps and systems, including Uber, Airbnb, and Upwork.com are mostly mash-ups of existing digital components.

The four digitech laws have made things that were unthinkable into things that are universal. They have opened doors to technologies that many thought could only come true in science fiction movies. But will this continue?

WILL THE DIGITECH IMPULSE CONTINUE?

The key to Moore's law up till now has been to cram more electronics on a single computer chip. Because things can only get so small, the end of Moore's law is a logical inevitability. Indeed, some think we have already reached the limit. Peter Bright of Ars Technica, for example, wrote in a November 2016 article, "Moore's law has died at the age of 51 after an extended illness."[4] Intel chief executive Brian Krzanich has a different view (as you might expect from the executive running the company Gordon Moore founded).

In May 2017, Krzanich announced that the death of Moore's law had been postponed. "I've been in this industry for 34 years," said Krzanich, "and I've heard the death of Moore's law more times than anything else in

---

4. Peter Bright, "Moore's Law Really Is Dead This Time," *ArsTechnica.com*, November 2, 2016.

my career. And I'm here today to really show you and tell you that Moore's law is alive and well and flourishing. I believe Moore's law will be alive well beyond my career."[5]

The transistors in today's microprocessors are about 14 nanometers wide. To give you an idea of how small that is, bacteria are between 10,000 and 100,000 nanometers, and the average virus is 100 nanometers. Individual atoms are on the order of a tenth of a nanometer. When Krzanich told everyone to call off the funeral for Moore's law, he was announcing a chip that would have transistors that are 10 nanometers wide.

Obviously, you can divide 10 nanometers in half quite a few times before you reach the size of an atom, but at that scale the world becomes strange in the quantum physics sense of the word. In the 2015 *Technology Roadmap for Semiconductors* report, the main author, Paolo Gargini, writes, "even with super-aggressive efforts, we'll get to the 2–3-nanometre limit, where features are just 10 atoms across." At that scale, electron behavior is governed by quantum uncertainties that would make transistors hopelessly unreliable. Gargini guesses that this limit will be reached in the 2020s.

Physical limits, however, need not stop computers from getting faster, cheaper, and smaller.

The "More Moore" and "More Than Moore" Ways Forward

To date, the industry has pursued what Gargini calls the "more Moore" route, that is, increasing the density of components on a single semiconductor. But there are more ways to boost computer power than the "more Moore" route. Gargini points out that engineers are coming up with techniques such as going from 2D chips to 3D chips.

Another way forward is what Gargini calls "more than Moore" approach, which is to make chips that are optimized for specific tasks rather

5. Daniel Robinson, "Moore's Law Is Running Out—But Don't Panic," *ComputerWeekly.com*, November 19, 2017.

than jack-of-all-trade computing. By analogy, the more-Moore route is like making an athlete ever stronger, so she could win medals in every strength sport. The new approach is to train some athletes for the shot put and others for the discus. The Nvidia chip for machine learning is a good example of the more-than-Moore way forward, since it is specifically designed for machine learning.

The ultimate solution to physical limits is quantum computing, which draws on the weird and wonderful properties of quantum physics where one thing can be many things at the same time. This, which some think will get out of the labs and into the workplace in the 2020s, promises a quantum leap in computing power. Quantum computing, however, is a long way from having commercial applications—noting, of course, that "a long way" in the world of digital technology is ten years.

There are other ways to get around the physics that put a limit on the shoehorning of more transistors into a single chip. A common one is to substitute transmission for local processing muscle. This is the trick iPhones do with the digital assistant Siri. On many iPhones, Siri only works when the phone is connected to the internet. Your voice data is compressed, whizzed over to Apple's supercomputers in the cloud, and the answer is whizzed back to your iPhone for Siri to deliver in her smooth voice. And all that in seconds, or microseconds.

These various ways forward seem likely to keep digitech advancing at a breakneck pace for years to come.

One of the most important things that the four laws have made possible is a very curious technology that carries the seemingly self-contradicting label of "machine learning". We can see just how strange machine learning is by looking at how people interact with the things it has enabled.

MACHINE LEARNING—COMPUTING'S SECOND CONTINENTAL DIVIDE

Amanda Barnes has a new colleague named Poppy. This pair helps insurance brokers at Lloyds of London comply with financial regulations that were established after the 2008 financial crisis. New insurance policies have to be listed with a central registry, and this means creating and validating a so-called London premium advice note, or LPAN. It's almost routine—call it "knowledge assembly line" work.[6]

The insurance broker sends an email with information on the new policy. Then someone has to open it, extract the relevant information, validate it, and match it with additional data. The LPAN is then filled out, and the whole package is uploaded to the Insurers' Market Repository.

Barnes can get through five hundred LPANs in a few days. Poppy does it in a few hours. Poppy is part of the new digital workforce where the "digital" refers to the worker not the work. She is a white-collar robot where the "white collar" refers to the attire of the workers she is replacing not the clothing that the robot is wearing. Poppy is an example of a new form of artificial intelligence called robotic process automation (RPA) which draws on the new capacities created by machine learning.

Barnes views Poppy as a co-worker despite the fact that "she" is really just a piece of software. Indeed, it was Barnes who gave the software a name. Perhaps this naming stems from the fact that the software does exactly what Barnes used to do, and in exactly the same way. Or maybe it is because the RPA seems vulnerable—Poppy cannot handle the tricky cases. Those she has to hand off to Barnes.

This sort of personification is pretty common when it comes to software robots. Ann Manning, a worker at the business processing company Xchanging, for example, trained an RPA and then called it Henry. "He is programmed with 400 decisions, all from my brain, so he is part of my brain and I've given him a bit of human character," she explained.[7] When a Texas Mercedes dealership implemented a virtual assistant to respond to car queries and set up appointments, the human sales representatives called it Tiffany, and customers loved "her." Joseph Davis, internet director

---

6. See Leslie Willcocks, Mary Lacity, and Andrew Craig, "Robotic Process Automation at Xchanging," Outsourcing Unit Working Research Paper Series 15/03, London School of Economics and Political Science, June 2015.

7. Willcocks, Lacity, and Craig, "Robotic Process Automation at Xchanging."

at the dealership, claimed, with a touch of Texan bravado: "We've had one client show up with roses for her, and a couple others have tried to ask her out."[8]

There are important hints here. People don't give nicknames to their laptops, smartphones, or Excel programs. The practice of naming software robots is a message from the frontlines informing us that this automation is really something new. And the frontline workers are right. Computers can now "think" in ways they never could before.

Computers Shift from Obedience to Cognition

Machines recently crossed a second "continental divide." The first, which came in the 1970s, was from things to thoughts, as we saw. The second is from conscious thought processes to unconscious thought processes. Think of it as the reversal of Moravec's paradox.

AI-pioneer Hans Moravec wrote (in the late stone ages of AI, 1988 to be specific): "It is comparatively easy to make computers exhibit adult level performance on intelligence tests or playing checkers, and difficult or impossible to give them the skills of a one-year-old when it comes to perception and mobility." That was the paradox; computers were good at what humans found hard, and bad at what humans found easy. This division reflected a feature of human thinking that has long been known to specialists.

Psychologists tell us that humans have two very distinct ways of thinking—conscious, careful, logical, verbal thought, on one hand, and unconscious, quick, instinctive, nonverbal thought, on the other. When you mentally calculate a 15 percent tip, you are using the logical way of thinking; instinct has nothing to do with it. When you catch your balance after stumbling, you are using the instinctive way of thinking; logic has nothing to do with it.

8. Quoted in Jesse Scardina, "Conversica Cloud AI Software Tackles Sales Leads," *TechTarget. com* (blog), June 1, 2016.

Being scientists, psychologists handed out less-than-poetic names for these two ways of thinking: System 1 (intuitive or instinctive thinking), and System 2 (analytic thinking). Social scientists have invented flashier names. The psychologist Daniel Kahneman, who won the 2002 Nobel Prize in Economics, called the two systems "thinking fast" and "thinking slow" in his 2011 book *Thinking Fast and Slow*. I prefer the terminology of New York University social psychologist Jon Haidt, who labels the slow-thinking, rational part of our brain as "the rider" and the fast-thinking, instinctive part as "the elephant."

Haidt's labels evoke the image of a small rider (who is an analytic, conscious thinker of the System 2 type) sitting atop a giant elephant (who is an instinctive, unconscious thinker of the System 1). Two aspects of this labeling are insightful (in a System 2 sort of way). First, the elephant does most of our thinking, even if we are unaware of it; the elephant does the heavy lifting when it comes to cognition. Second, although the rider sits atop the elephant and is, in principle, in control, the reality of who controls whom is less clear than it seems. It is very hard for the rider to control the elephant—as anyone who has vowed to lose weight can attest.

But what has this got to do with digital technology and RPA like Poppy? The deep source of Moravec's paradox was the nature of traditional computer programming. Traditional programming mimicked the way the rider thinks, not the way the elephant thinks.

Until a few years ago, we humans taught computers to do things with computer programs.[9] These programs explained, step by logical step, what the computer should do in every possible situation it might encounter. But this approach meant that before we could teach computers to think, we had to understand how we think, step-by-step.

Moravec's paradox arose since, as another early hero of AI, Marvin Minsky, put it, "we're least aware of what our minds do best." We understand how our rider thinks—how we, for example, do arithmetic, algebra,

9. Machine learning has been around for decades, but a lack of computer power and data limited the effectiveness of the algorithms it produced in the past.

and archery. We haven't a clue as to how our elephant thinks—how we, for example, recognize a cat or keep our balance when running over hill and dale. A form of AI called "machine learning" solved the paradox by changing the way computers are programmed.

With machine learning, humans help the computer (the "machine" part) estimate a very large statistical model that the computer then uses to guess the solution to a particular problem (the "learning" part). Thanks to mind-blowing advances in computing power and access to hallucinatory amounts of data, white-collar robots trained by machine learning routinely achieve human-level performance on specific guessing tasks, like recognizing speech. With machine-learning-trained algorithms, computers started to think, to cognate. It was no longer a case of computers just following explicit instructions. They now can make educated guesses in ways that are giving them the ability to undertake some forms of human thinking. And that's why machine learning is affecting the world of work in such radically new ways.

This new form of computer cognition is changing realities. It is creating new forms of automation that will replace millions of humans whose jobs were—until the twenty-first century—sheltered by the fact that computers couldn't handle elephant/think-fast/System 1 tasks. Now they can. Machine learning is really a revolution that everyone needs to understand. It has made headlines when it comes to game-playing, so that's a good place to start.

Games and Beyond

The ancient board game 'Go' is way more complex than chess. After two moves, a chess player has 400 possible next moves. After two moves in Go, a player has 130,000 possible moves—and it just gets more complex. There are more possible positions on a Go board than there are atoms in the universe. The game is so complex that the best human players instinctively "sense" what to do. They cannot, as in chess, puzzle through their strategy in a logical fashion.

This complexity is also why computers using rider/think-slow/System-2 "thinking" couldn't match human-level performance in Go even though they beat the best humans at chess decades ago. That changed in May 2017. That's when a computer program, called AlphaGo Master, used machine learning techniques to beat the world's best Go player.[10] The how is as amazing as the what.

AlphaGo Master, owned by the leading AI company DeepMind, learned the ropes by studying 30 million board positions from 160,000 actual games. This is a bit intimidating. There are only about 26 million minutes in a human working life, so AlphaGo Master started with more than a lifetime of experience. But then things got even more daunting for human players hoping to compete with this technology.

To learn from experience, AlphaGo Master played more games against itself in six months than a human could play in six decades. As Ke Jie, the world's best player put it after he lost to the algorithm: "Last year, it was still quite human-like when it played. But this year, it became like a god of Go." But that's not the end of the amazing part.

In a classic example of AI's inhuman speed, the owner of AlphaGo Master developed a new version of AlphaGo that skipped the "learning from human games" part and just let it learn from playing itself from scratch. All it started with were the rules. Since computing power had increased so much since AlphaGo Master was "trained," the results were astounding. In just 40 days of playing itself, the new version, AlphaGo Zero, beat the world's best Go player, which, at the time was AlphaGo Master. The victory came just six months after AlphaGo Master's astounding victory over the best human player.

But machine learning is not just fun and games. Computer scientist are pushing beyond headline-grabbing game playing to job-grabbing automation. Before machines crossed the second continental divide with machine learning, computers were not very good at office work. They couldn't read handwriting, recognize people, write, speak, or understand speech. Now they can—and their office skills are getting better fast.

10. Elizabeth Gibney, "Self-Taught AI Is Best Yet at Strategy Game Go," *Nature*, October 18, 2017.

One example provides an excellent way to understand what machine learning is, how it works, and how it is limited. The example is the way Siri "learns" a new language, in this case the Shanghai dialect of Chinese, called Shanghainese.[11] While one of the key ingredients is massive computer power, this example starts with a great deal of human effort.

How Siri Learned Shanghainese

Apple computer scientists got Shanghainese speakers to read out sample words and paragraphs. This created a database where particular sounds (speech) are linked to particular words (text). This is called the "training data set."

Computers can't hear in the human sense; they can deal only with inputs that have been digitized—that is, turned into strings of zeros and ones. That's why the sound and the text had to be "digitized." The recorded sound waves are translated into strings of zeros and ones, as are the words they correspond to in the training data set. This yields a computer-readable data set in which one pattern of zeros and ones (speech) is known to correspond to another pattern of zeros and ones (text). This is where machine learning steps in.

The chore is to identify which features of the digitalized speech data are most useful when making an educated guess as to the corresponding word. To tackle this chore, the computer scientists set up a "blank slate" statistical model. It is a blank slate in the sense that every feature of the speech data is allowed to be, in principle, an important feature in the guessing process. What they are looking for is how to weight each aspect of the speech data when trying to find the word it is associated with.

The revolutionary thing about machine learning is that the scientists don't fill in the blanks. They don't write down the weights in the statistical model. Instead, they write a set of step-by-step instructions for how the

11. See the fascinating description of the process by Benjamin Moyo, "Apple Speech Team Head Explains How Siri Learns a New Language," *9to5Mac* (blog), March 9, 2017.

computer should fill in the blanks itself. The human-written instructions tell the machine how to learn about which features of the sound data are the important ones. Putting it differently, the scientists "teach" the computer how to "learn" what the best weights are by studying the pairings in the training data set.

These human-written instructions tell the computer to be bold at first—to just make wild guesses about the weights. Think of this as a rough first pass. The computer then gives itself a pop quiz to test the accuracy of the rough-first-pass guesses. After grading its own pop quiz, the computer jiggles the weights to see if it can improve its score on the next pop quiz. By playing around with the weights, going back and forth between the weights and pop quizzes, the computer eventually arrives at what it considers to be a really good set of weights. That is to say, it identifies the features of the speech data that are useful in predicting the corresponding words.

The scientists then make the statistical model take an exam. They feed it a fresh set of spoken words and ask it to predict the written words that they correspond to. This is called the "testing data set." Usually, the model—which is also called an "algorithm"—is not good enough to be released "into the wild," so the computer scientists do some sophisticated trial and error of their own by manually tweaking the computer program that is used to choose the weights. After what can be a long sequence of iterations like this, and after the statistical model has achieved a sufficiently high degree of accuracy, the new language model graduates to the next level.

Apple didn't immediately use this new algorithm for translation. It used it to generate even more data. The new language algorithm was released as a new option on Apple's iOS and macOS dictation feature (the thing that fires up when you touch the microphone icon that is next to the spacebar on your iPhone keyboard). As native Shanghainese speakers used the feature, Apple recorded speech samples. It then had humans map these into text to create a new training data set of paired sounds and text. The computer was then sent back into the classroom for a few more thousand or a few more million rounds of weight-jiggling and pop-quizzing. This back and forth continued until Apple was satisfied with the statistical model's performance.

This is what lets Siri "understand" a new language. Learning to "speak" is a lot less clever. Human actors record lots of words and speech sequences in Shanghainese for Siri to play back to humans in reply to various queries and requests.

AI has been around for decades—the term was coined in 1956. And even machine learning is old hat, so the question is: Why now? Why did machine learning get so good so fast?

Why Machine Learning Now?

The easy answer lies in just two words—computing power—or maybe four words: much more computing power. It's Moore's law in operation.

Training AI systems to recognize photos or understand spoken language at human levels requires astounding amounts of computer horsepower. To get technical, the weight-jiggling part of machine learning involves a mathematical operation called "matrix inversion." Doing this for large systems involves an unbelievably large number of calculations. For an algorithm that is looking at, say, hundreds of thousands of pixels, a single inversion involves millions of billions of calculations.[12] That, in turn, is only feasible with processing power that used to be unthinkable for anything but the fastest supercomputers. Moore's law removed that limitation. Computer speeds that were out of reach in 2014 became run-of-the-mill in 2016.

The other reason this is happening now is that it is possible to collect, store, and transmit big data sets.

Fast computing and big data are linked for a very simple reason. If computer capacity is machine learning's jet engine, data is the jet fuel. While Moore's law cranked up the engine power, Gilder's law kept the fuel pumping. The size of the data sets being used is something that was thinkable but not doable just a few years ago. Big data today can get gigantic.

---

12. The computational complexity of inverting an n by n matrix is on the order of n cubed.

The website Flickr, for example, posts 100 million videos and images that can be used for training image recognition algorithms. To think about how big that is, note that it takes about seventeen minutes to count to a thousand by ones. Taking a break now and again, you could count up to three thousand in an hour. Doing that forty hours a week, fifty weeks a year would get you up to six million in a calendar year. You'd need another sixteen years or so to get up to a hundred million—and by then, Flickr probably would have doubled their dataset size several times.

But with all this amazing computer power and all these big data sets, why don't we see machine learning deployed more widely? One problem is that once AI gets good enough, we stop thinking of it as AI. For example, Optical Character Recognition, which lets you scan a document and turn it into a Word file is AI, but most people just think of it as a standard feature. In other words, we already are surrounded by AI, but we don't know it. A second problem is a skill shortage.

RPA systems like Poppy or Henry can be trained very easily by people with only minimal training in the training. But getting high-end AI systems to work is a very different proposition. It requires people with advanced education and lots of experience. As it turns out, there just aren't enough AI scientists to turn the possibilities into a real-world revolution. By some estimates, only ten thousand people worldwide have what it takes to build complex AI systems like Amelia, Siri, or Cortana. Google, however, has a solution.

Google has developed a set of tools that reduces the need for high-skilled human input into machine learning. Released in January 2018, it is called AutoML, short for "automated machine learning." This is really the stuff of Sci-Fi. AutoML is a machine-learning program that is learning how to design machine-learning algorithms on its own. It is like a robot building other robots, or at least a robot helping humans build robots. The goal, according to Google, is to allow hundreds of thousands of programmers who are good but not geniuses to develop new machine-learning applications. Today, many companies in many service sectors have vast data sets, but they can't exploit them without AI systems trained with machine-learning techniques. AutoML will accelerate service-sector automation by alleviating this constraint.

While machine learning allows computers to complete many human-like mental tasks, the outcome is far from human-like thinking. There is a lot of confusion on this point due in part to the fact that machine learning is called "artificial intelligence"—a phrase that seems designed to confuse.

AI as "Almost Intelligent"

Names can cause confusion. "Artificial Iintelligence" is a prime example. Everyone is absolutely certain they know what "intelligence" means and what "artificial" means. Put the two words together and we get confusion and misunderstanding rolled into an ominous sense that can border on fear. Or maybe we get scoffing and laughter. "Artificial Intelligence" is not a phrase that rings the same bells for everyone.

Some of us think of goofy science fiction characters like C3PO in *Star Wars* or the robot maid Rosie-the-maid in the 1960s TV show *The Jetsons*. Others think of terrifying characters like the unstoppable, silver-liquid T-1000 in *The Terminator* movie, the psychopathic computer "Hal" in the movie *2001: A Space Odyssey*, or the computer manipulating humans in *The Matrix*.

The easy definition of artificial intelligence is a computer program that can "think" and thus has some form of intelligence. But what then is intelligence? Psychologists define intelligence as: "A very general mental capability that, among other things, involves the ability to reason, plan, solve problems, think abstractly, comprehend complex ideas, learn quickly and learn from experience."[13] Today's AI is not intelligent in this sense.

Machine learning does only the last two items in the psychologists' list: learn quickly and learn from experience. Even the revolutionary machine learning applications we see today—like Siri and self-driving

---

13. Linda Gottfredson, "Mainstream Science on Intelligence: An Editorial with 52 Signatories, History, and Bibliography," *Intelligence* 24, no. 1 (1997).

cars—are just computer programs that recognize patterns in data and then act, or make suggestions based on the patterns they find. The pattern recognition is astonishing, often superhuman in specific areas. But pattern recognition is not "intelligence" as the word is generally used when speaking about intelligent animals like humans, chimpanzees, or dolphins. AI should really stand for "almost intelligent," not artificial intelligence.

Digital technology is an amazing thing to behold. To some it is fascinating. To others it is frightening. But one thing that should be obvious to all is that it will change our economies, our lives, and our communities.

FROM TECHNICAL IMPULSE TO ECONOMIC TRANSFORMATION

As we have seen, digital technology has launched a new four-step progression: transformation, upheaval, backlash, and resolution. The first step—economic transformation—is already underway and is driven by the familiar dynamic-duo of economic change: automation and globalization.

The Globotics Transformation differs from the earlier ones in two important ways. The first is size. Digitech's impact will be felt most heavily in the service sector. Since most people work in the service sector, the impact on societies will be much greater than the Service Transformation, which mostly disrupted the manufacturing sector. Even at the height of manufacturing's importance, less than a third of workers had jobs in this sector, so the social impacts, while traumatic, were limited to a relatively small share of workers. This time, the impact will be much more broadly felt.

The second big difference is the timing. Unlike the transformation we experienced in the nineteenth and twentieth centuries, both members of the dynamic duo—automation and globalization—are swinging into action at the same time. That is what puts the "globotics" in this book's title. We need to stop asking whether the economic impact is due mostly to globalization or mostly to automation. Globalization and robotics are now Siamese twins—driven by the same technology and at the same pace.

In the past two transformations, the technological impulses launched new forms of automation long before they launched new forms of globalization. To emphasize the fact that digitech fired the starting gun for white-collar globalization and white-collar automation at the same time, we look at globalization first.

5

Telemigration and the Globotics Transformation

Mike Scanlin is a restless soul. With three careers behind him (software engineering, investment banking, and venture capital), he decided to move to Las Vegas and follow his passion. This, oddly enough, is "covered options"—a fiddly investment strategy that involves selling stock options on stocks that one owns.

Actually, covered options are only passion number two for Scanlin. "My passion and #1 hobby is travel and hiking," he related, but "I was never off-line for more than about 36 hours (yes, you can get a cell signal from Base Camp Everest; helps if you're on a ridge and not in a valley)."[1]

To get his start-up to the point where the pinnacle of Machu Picchu, the bottom of the Grand Canyon, and the Zion Narrows River hike were possible workplaces, he hired talented professionals based abroad. He spent $37,000 on help from IT engineers and web designers that he figures would have cost him $500,000 if he had hired in America. Now, he just can't imagine a time when he won't use online foreign freelancers to get projects done.

1. Quoted in Camila Souza, "41 Entrepreneurs Share Their Unusual Hobbies," *Tech.co* (blog), May 21, 2015; also see TJ McCue, "3 Freelance Economy Success Stories," *Forbes.com*.

Such practices have not yet attracted much attention from the wider public, but they should have and probably soon will. They really are a big deal. The choices made by people like Scanlin are bringing American and European office workers into direct wage competition with talented, foreign workers willing to work for little money.

Of course, the internet is a two-way street and wage competition isn't always won by the cheapest. That's why international freelancing is also creating new opportunities for some advanced-economy workers. Firms often hire more expensive, more experienced workers when they need something done right. This is why service companies from high-wage nations have long dominated world markets in sectors like finance, accounting, engineering, telecommunications, and logistics. Their competitive advantage is based on excellence, not low wages.

But whichever side of the street you are on, this is really something different. Before today's digital wonders, the only way Scanlin could have hired foreign programmers was if they had immigrated to the US. In that case, they surely would have demanded wages and benefits in line with US standards.

INTERNATIONAL WAGE COMPETITION FROM TELEMIGRANTS

What these foreign online workers are doing—in a virtual sense—is migrating temporarily into Scanlin's company and working at wages that make sense in their home countries. And those wages are often very low. Salaries in the US and Europe are typically a dozen times what they are in developing nations.

Figure 5.1 shows that an accountant in China earns about one-twentieth of the salary of a US accountant. The Chinese accountant would be unable to do all, or even most, of a US accountant's job, but at twenty times cheaper, there are some tasks the Chinese accountant could take over from high-priced US accountants. With help provided by the Chinese assistants to US accountants, US firms could get through the work stack with fewer

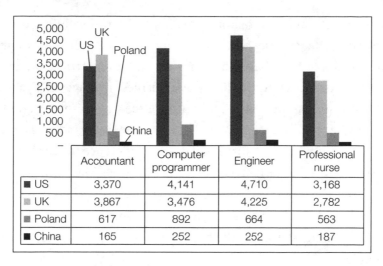

	Accountant	Computer programmer	Engineer	Professional nurse
■ US	3,370	4,141	4,710	3,168
▨ UK	3,867	3,476	4,225	2,782
■ Poland	617	892	664	563
■ China	165	252	252	187

Figure 5.1 How Much Cheaper Are Foreign Workers? Net Monthly Income, in 2005 US Dollars.
SOURCE: Author's elaboration of ILO online data: Net Monthly Income (constant 2005 US dollars).

locals. For example, instead of employing ten US accountants, a company could get the job done at a much lower cost with seven local accountants and seven remote assistants. And it might end up doing a better job. By paying a bit more than the average Chinese salary, a US firm could get the cream of the crop among Chinese accountants—the most clever and diligent ones. This would mean that the ground work would be done by top-notch foreign workers instead of second-rate locals.

The cost savings are similar for computer programmers, engineers, and nurses. In each case, complete replacement would be impossible, but some substitution of low-cost foreign workers for high-cost domestic workers would obviously save money.

It worked for me. In April 2018, I hired a copyeditor sitting in Bangkok to go over blog posts for the policy portal that I run in London (VoxEU. org). She has a masters degree in International Relations from Columbia University and a very sharp eye for errors made by my authors—many of whom are non-native speakers. At $25 per hour, she is about 35% cheaper than the European copyeditors I use. But affordable service workers are not only accessible to companies.

You yourself can hire a remote personal assistant for little money. For example, one site, avirtual.co.uk, lists Leigh McLaren-Brierley as an online personal assistant. Based in Cape Town, South Africa, she is a native English speaker with experience as a business manager at Thompsons Travel and a special interest in human resources, recruitment, and travel planning. The peppy quote next to her get-to-know-you video says: "I love what I do because I believe that I make a real difference to my client's productivity and life." Alternatively, there is Monique Mancilla, who has a BA from the University of Santa Barbara and experience in bookkeeping and social media; she speaks English and Spanish fluently. At avirtual.co.uk, 270 pounds sterling is the basic rate for fifteen hours of assistance per month.

While there is little systematic data on the rates charged by freelancers around the world, some survey evidenced exists. One was done by a new freelance matchmaking site: freelancing.ph. The site was set up to help Filipinos establish careers online. As their marketing material says, "We believe that with the right mindset Filipinos can unleash their world-class potential." To help promote telemigration, the site conducted a survey of how much their freelancers earned. Remembering that these shockingly low rates are meant to attract Filipinos to the site, the survey results are very revealing. Workers in the job category "digital marketing strategists" earned between $6 and $8 an hour, general virtual assistants got between $3 and $8, content editors and financial managers came in at about $6 to $15.[2]

Although that sounds like little money in the US or Europe ($10 an hour translates into an annual income of $20,000), it is above average in most of these countries. In the Philippines, for example, the national average income is $9,400. The World Bank did a study of international freelancing where they found that full-time online workers in Kenya, Nigeria, and India make more than their peers who have traditional jobs.

In this sense, telemigration, or international telecommuting is win-win for the companies and the freelancers. My website, VoxEU.org is saving

2. See "2016 Pinoy Freelancer Salary Guide," on *freelancing.ph.*

money, and my Bangkok-based copyeditor is earning more than she would locally. The only ones who may be less-than-happy about this arrangement are the European copyeditors who are getting less work.

Low wages are not the only advantage of foreign freelancers—they also offer access to a much deeper pool of talent. Moreover the emergence of new matchmaking platforms is making it easy to find, hire, manage, pay, and fire telemigrants. That's what the CEO of ThePatchery.com found out.

ONLINE MATCHMAKING PLATFORMS

Amber Gunn Thomas had a brainwave. She loved sewing clothes for her kids and thought, why not make a business out of it? Why not let people design clothes for their own kids? To set up the website for her new business (ThePatchery.com), she hired a local web design company in Minnesota. They burned through her development budget before the job was really done, so she turned to foreign online workers. But how did she find foreign workers while sitting in Minnesota?

The answer is that she used an online matchingmaking platform. These web-based matchmaking platforms are very much like eBay, but for services rather than goods. eBay helps people and companies buy and sell goods online. These freelancing sites help people and companies buy and sell services online.

After interviewing a few freelancers online, Gunn Thomas hired a Belarusian agency, iKantam. "It changed the course of our business," she said. The work was done faster than the local agency, and iKantam brought a level of expertise that Thomas had not seen with the local web development company.

Hiring remote foreign workers is not just for small companies like ThePatchery.com. Big companies are embracing it too. American Express, for example, is turning to foreign freelancers for many jobs. "Having a remote workforce allows us to cast a wider net, reaching prospective employees who may not live within commuting distance of one of our brick-and-mortar customer care locations," is how Victor Ingalls, vice president of world service at American Express, explains it. He also explains that having people in different time zones helps the company deal with customer demands during off-hours. It is also helpful that remote workers are willing, often eager, to work part time or on nontraditional schedules.[3]

Many other corporate giants post help-wanted ads on freelance sites. On Flexjobs.com, you can find listings for telecommuting posts in engineering and architecture from Dell and Deloitte, or remote project-management jobs with Xerox, UnitedHealth Group, and Oracle; communications jobs with CBS Radio; and travel and hospitality jobs with Hilton. The list goes on and on.

Foreign freelancers also offer extreme flexibility. Thanks to the freelancing platforms, they are easy to find, hire, manage, and fire—a feature which is a big draw for employers.

Finding, Hiring, and Managing Foreign Workers

The world's largest online site for matching workers and projects is called Upwork.com—that's where I hired my copyeditor. I wrote up a description of the work I wanted done and the qualifications of the freelancer I wanted to hire. This went up on the site as a "job posting" that freelancers could respond to it with "proposals." What I got was a dozen or so proposals, including some from freelancers that were suggested by Upwork's matchmaker bot.

After reading the proposals (short cover letters) and checking out their online profiles (which included the wage they were asking), I interviewed two of them online for about 15 minutes each. After hiring my preferred candidate, I started posting work via Upwork's file sharing service, and communicating with the copyeditor on the site (the site sends me an email when there is a new message, or file posted). To reassure me that the hours

---

3. "Another 10 Companies Winning at Remote Work," *CloudPeeps* (blog), May 17, 2016.

billed by the freelancers are real, Upwork takes occasional screenshots of the freelancer's screen while she is claiming to be working for me.

The freelancer knows she'll get paid since Upwork automatically charges the credit card I posted. I can object to the billing if something goes wrong or the work is substandard, but so far so good. We both have an interest in making it work since it's win-win. If something did go wrong, the work dried up, or I decided to switch to another freelancer, firing a freelancer is simplicity itself. You click on a a button labelled "End Contract".

I am mostly definitely not the only one doing this. In 2017, Upwork had fourteen million users from over 100 nations. It processed more than one billion dollars in freelancer earnings. And Upwork has plenty of competition. There are dozens of start-up competitors like TaskRabbit, Fiverr, Craigslist, Guru, Mechanical Turk, PeoplePerHour, and Freelancer.com. This "space," as they say in the online world, has attracted the attention of the professional network giant LinkedIn. It has 450 million business professionals registered and it is using that base to move into freelance matchmaking with its "ProFinder" services. And then there is the Chinese entrant.

As you might expect given how digital the Chinese economy has become, online freelancing is booming in China. Zhubajie (zbj.com) is the largest platform. It started in 2006 and now has more than sixteen million freelancers registered. More than six million businesses have used its network. The company is also expanding internationally. Its English-language portal, Witmart.com, caters to customers globally.

The CEO Zhu Mingyue explains this new form of globalization will be more sudden than traditional globalization: "Compared with online goods trade, our services trade has no constraints in terms of logistics and customs. It is very promising." The company has already set up offices in Houston in the US and Toronto in Canada. "We are based in China and mainly serve the Chinese clients, but we aim at the global market."[4]

---

4. He Huifeng, "Zhubajie Charges toward Unicorn Status, and Flotation," *South China Morning Post*, July 1, 2016.

I think it is very likely that other emerging markets will set up their own matchmaking platforms to help their citizens join the world of international freelancing. It would be an excellent way for them to create jobs for their rapidly expanding workforces.

In a sense, these web platforms are affecting telemigration in the same way that railroads, containers ships, and air cargo affected trade in goods. By radically lowering the cost of moving goods internationally, better transportation technology allowed companies to exploit international goods-price differences. The result was booming trade in goods. By radically lowering the cost of hiring foreign service workers, freelance platforms are allowing companies to exploit international wage differences. The result will surely be an explosion in telemigration.

Who Are These Foreign Freelancers?

Given the unconventional nature of this work, official statistics tend to be absent or misleading. To fill in some of the blanks, Oxford professor Vili Lehdonvirta has setup an innovative project to track online labour—the iLabour Project.[5] He finds that almost a quarter of online freelancers are working from India, and another quarter are based in Bangladesh and Pakistan. The other big emerging-market supplier is the Philippines, but fully an eighth are from the UK and the US.

Another glimpse into the world of freelancers comes from a large-scale survey that focused on freelancers from low-wage nations (done by the online payments company Payoneer.com). They queried twenty-three thousand freelancers worldwide. About a quarter of respondents were in Latin America and Asia, twenty percent in Central and Eastern Europe, and about fifteen percent in both the Mideast and Africa.[6]

5. "The iLabour Project, Investigating the Construction of Labour Markets, Institutions and Movements on the Internet", ilabour.oii.ox.ac.uk. Also see "Digital Labor Markets and Global Talent Flows" by John Horton, William R. Kerr, and Christopher Stanton, NBER Working Paper 23398, April 2017.

6. Melisa Sukman, The Payoneer Freelancer Income Survey 2015.

The vast majority of freelancers surveyed are in their twenties and thirties. A bit more than half had university educations. The companies paying for their services were about half in North America and Europe (split equally), about fifteen percent in both Latin America and Asia, and seven percent in Australia and New Zealand.

Looking at the list of countries where telemigrants are coming from makes it clear that language is a big issue in digitally-enabled globalization of service and professional jobs. This makes perfect sense. Services are personal in a way that goods are not. It makes no difference, for example, that you cannot talk with the person who helped you by assembling your iPhone. It makes a huge difference if you cannot talk with the person who is helping you with your travel arrangements.

The fact that most freelancing jobs require "good enough" English has greatly restricted the pool of potential telemigrants. Digital technology, however, is relaxing that restriction thanks to an amazing application of AI called "machine translation." Instant translation used to be the stuff of science fiction. Today it is a reality and available for free on smartphones, tablets, and laptops. It is a long way from perfect, but progress since 2017 has been absolutely amazing—as a French tourist in Iceland found out in 2017.

MACHINE TRANSLATION AND THE TALENT TSUNAMI

In August 2017, an Icelandic landowner caught a French tourist fishing illegally on his land and called the police. Once the tourist worked out that the police were on their way, he seemed to lose his mastery of English. But that didn't slow the course of justice. Not in today's world. The officer interrogated him with Google Translate and gave him a big fine as a souvenir of his fishing expedition.

In the same month, a UK court used Google Translate because someone forgot to arrange for a human translator for the Mandarin-speaking defendant. The defendant was happy to proceed without a human translator since Google Translate is now so accurate. In June 2017, the US Army paid

Raytheon four million dollars for a machine translation package that lets soldiers converse with Iraqi Arabic and Pashto speakers as well as read foreign-language documents and digital media on their smartphones and laptops.

Machine translation used to be a joke. A famous example, related by Google's director of research Peter Norvig, was what old-school machine translators did with the phrase, "the spirit is willing but the flesh is weak." Translated into Russian and then back to English, it turned into "the vodka is good but the meat is rotten."[7] Even as recently as 2015, it was little more than a party trick, or a very rough first draft. But no longer. Now it is rivaling average human translation for popular language pairs.

According to Google, which uses humans to score machine translations on a scale from zero (complete nonsense) to six (perfect), the AI-trained algorithm "Google Translate" got a grade of 3.6 in 2015—far worse than the average human translator, who gets scores around 5.1. In 2016, Google Translate hits numbers like 5.[8] And the capabilities are advancing in leaps and bounds.

As is true of almost everything globots do, machine translation is not as good as expert humans, but it is a whole lot cheaper and a whole lot more convenient. Expert human translators, in particular, are quick to heap scorn on the talents of machine translation.

The Atlantic Monthly, for instance, published an article in 2018 by Douglas Hofstadter doing just this.[9] Hofsadter is a very sophisticated observer with very high standards when it comes to machine translation. With a father who won the 1961 Nobel Prize in Physics, a PhD in physics to his name and now a post as a professor of cognitive science, he is someone who knows what he is talking about. As he puts it: "The practical utility of

7. Stuart Russell and Peter Norvig (2003). *Artificial Intelligence: A Modern Approach* (Englewood Cliffs, NJ: Prentice Hall, 2003).

8. Yonghui Wu et al., "Google's Neural Machine Translation System: Bridging the Gap between Human and Machine Translation," *Technical Report*, 2016.

9. Douglas Hofstadter, "The Shallowness of Google Translate," *The Atlantic Monthly*, January 30, 2018.

Google Translate and similar technologies is undeniable, and probably it's a good thing overall, but there is still something deeply lacking in the approach, which is conveyed by a single word: understanding." But then he goes on to reveal a deep abhorrence of machine translation.

Writing about the day when AI gets so good that human translators become mere quality checkers, he states that this would "cause a soul-shattering upheaval in my mental life. . . . the idea frightens and revolts me. To my mind, translation is an incredibly subtle art that draws constantly on one's many years of experience in life, and on one's creative imagination." Translation may be a subtle art to Hofstadter, but to most businesses struggling to do business internationally, translation is just a tool. Good-enough translations are, well, usually good enough.

Another skeptical professional translator made a similar point in The Independent newspaper in 2018. The author, Andy Martin, is a lecturer at Cambridge University. He teaches students how to translate French literary texts even though this is basically impossible. "It's like paying someone to teach tight-rope walking who assumes you are just naturally going to get blown off in a high wind or slip and fall into a void of pure nonsense," he writes. While he is willing to concede that machine translation is functional, he denies it could ever replace real humans completely: "Google is often adequate . . . but only in the way of a particularly uninspired apprentice translator." Real translation is not a matter of big data and algorithms; it's a subject for art: "There is, at the core of the translation process, a mystery, an almost mystic transcendence. There is no direct equivalence of one language to another."[10]

What this suggests is that the high-end translation is likely to stay in the hands of humans, but in the meantime international business will be transformed when these uninspired apprentice translators massively lower, but don't eliminate, language barriers.

Instant, free machine translation is not something that is lurking in computer laboratories. Free apps like Google Translate and iTranslate

10. Andy Martin, "Google Translate Will Never Outsmart the Human Mind," *The Independent*, February 22. 2018.

Voice are now quite good across the major language pairs. Other smart-phone apps include SayHi and WayGo. And machine translation is widely used. Google, for example, does a billion translations a day for online users.

Try it out. Machine translation works on any smartphone. Just open up a foreign language website and apply Google Translate to the text. You can even use the iTranslate app to instantly translate a foreign language in real time. You fire up the app on your smartphone and point your phone's camera at a page of, say, French, and you see the English translation on your phone's screen. Instant and free.

YouTube has instant machine translation for many foreign-language YouTube videos. You just go to the settings "gear," click on captions, and choose "auto-caption." Instant, free spoken translation is also possible with the add-on option Skype Translator. This will allow you to under-stand foreign-language speakers you are Skyping with. It is not perfect, but being able to Skype freely with someone who doesn't speak your lan-guage is nothing short of marvelous.

Microsoft and Amazon have entered the race as well. Microsoft is using its digital assistant, Cortana, to allow users to speak in any of twenty lan-guages and have the results appear as text in up to sixty different languages. Its email app, Outlook, added an instant translation add-in in 2018. At the end of 2017, Amazon introduced its contender—Amazon Translate—via Amazon Web Services.

Unbuilding the Tower of Babel

The fact that machine translation is entering everyday life is a big change. As anyone who has traveled or done business internationally knows, lan-guage is a huge barrier to just about everything. There is even an Old Testament story that says language-linked divisiveness was divinely inspired.

The passage, from the Book of Genesis, discusses a building that humans were constructing to reach the heavens: "The Lord said, 'If as one people speaking the same language they have begun to do this, then nothing they plan to do will be impossible for them. Come, let us go down and confuse their language so they will not understand each other.'" The structure came to be known as the Tower of Babel, where "babel" means a confused noise made by a number of voices. Not to put too fine an edge on it, machine translation is unbuilding the Tower of Babel. This, in turn, is accelerating the pace at which American and European office workers are coming into direct competition with talented, low-cost workers sitting abroad.

Of the 7.2 billion humans, about 400 million speak English as their first language. Adding in a generous estimate of non-native English speakers brings the number up to about a billion English speakers. Although there is some online freelancing in other major languages, English dominates the market to date, so only a billion people are potential participants in the new online freelancing movement.

With machine translation being so good, and getting better so fast, the billion who speak English will soon find themselves in much more direct competition with the other six billion who don't. Think about that. Then think about it again.

Machine translation means that all this foreign talent soon will speak English or other rich-nation languages like French, German, Japanese, or Spanish—not perfectly, but well enough to telemigrate for some jobs. The result will be a tsunami of global talent. All around the world, special people will find themselves suddenly less special.

Focus on China for example. Since about 2001, China has produced more university graduates than the US. The number now is over 8 million graduates per year. Only 8 percent of these Chinese graduates are unemployed, according to Katherine Stapleton, a researcher at the University of Oxford, but most are underemployed. They find work, but it is often part time or involves low-paid jobs for which a degree is not really necessary. Six months after graduating, a quarter of Chinese university degree holders earn less than the average Chinese internal migrant worker. The high living costs in China's big cities have, according to Stapleton, "forced millions of graduates into 'ant tribes' of urban workers living in squalid conditions—often in basements—working long hours in low-paid jobs."[11] Ant tribes sounds harsh, but that is the literal translation of the term used in China.

Just imagine the increase in competition that will happen now that these "ant tribes" can speak good-enough English (via machine translation) and sell their brain power over the internet to the US, Europe, Japan, and other rich nations.

But why is this only happening now? The deep answer is Moore's law and Gilder's law have shifted into their eruptive growth phases when it comes to machine translation.

WHY NOW? THE DEEP LEARNING TAKEOVER

For a decade, hundreds of Google engineers made incremental progress on translation using the traditional, hands-on approach. In February 2016, Google's AI maharishi, Jeff Dean, turned the Google Translate team on to Google's homegrown machine-learning technique called Deep Learning.

The job required huge amounts of computer muscle, but Google had that thanks to Moore's law. The missing link was the data. That changed in 2016 when the United Nations (UN) posted online a data set with nearly 800,000 documents that had been manually translated into the six official UN languages: Arabic, English, Spanish, French, Russian, and Chinese.

It is worth reflecting for a moment on how difficult it would have been to create, store, and upload that much data just a few years ago. It wasn't so long ago that downloading a feature-length movie was a task that strained most people's internet connections. It was Gilder's law that changed that reality, and today it is allowing the waterfall of language data to continue flowing.

For example, the EU Joint Research Center posted a dataset with human-translated sentences in twenty-two languages (it has over a billion words). Not to be outdone, the EU Parliament released a dataset with 1.3 billion paragraphs that had been translated into twenty-three

---

11. Katherine Stapleton, "Inside the World's Largest Higher Education Boom," *TheConversation. com,* April 10, 2017.

EU languages. Another massive database, uploaded by the Canadian Parliament, has millions of paired, human-translated sentences from the parliamentarian debates.

With data and the computer power to process it, Google translations improved more in a month than they had in the previous four years.[12] A couple weeks later, all projects using the old approach were halted. By fall 2016—just six months after the change—Google Translate switched fully to the new system. But they didn't tell anyone. They wanted someone else to tell the world about this revolution.

In November 2016, a Tokyo University professor of human-computer interaction, Jun Rekimoto, noticed Japanese to English translation had suddenly improved by an almost immeasurable amount. He sounded the alarm on his blog and Google then explained the changes at a press conference.

Almost as important are the rapid advances in communication technologies; these are making it seem almost as if foreign freelancers are sitting side-by-side with us even when they are in a different country. As with machine translation, this is no longer something only seen in *Star Trek* episodes, or the *Hitchhiker's Guide to the Galaxy*. What I like to call "Advanced Communications Technology For Acting Remotely" (ACTFAR) is a reality today.

COMMUNICATION TECHNOLOGY FOR MASS TELEMIGRATION

"You sit down at the table with your tablet and put on a pair of light-weight glasses. Suddenly the room comes to life. To your left, you see your colleague Jessica, who's joining from New York. To the right, the company CEO, Beth, who's currently in Atlanta. Across the table from you is Hassan, who's joining from his home office in London . . . . they're so

12. Gideon Lewis-Kraus, "The Great A.I. Awakening," *New York Times Magazine*, December 4, 2016.

lifelike it still startles you." This is the future vision of Stephane Kasriel, the Frenchman who runs Upwork.com.[13]

As it turns out, the kid-stuff technologies that have been revolutionary in the video-gaming world are about to have revolutionary impact on the world of telecommuting. The two key technologies are augmented reality (AR) and virtual reality (VR). Many companies, both start-ups and giants like IBM, are in the process of using AR and VR to improve remote collaboration. They are redefining what it means to work side by side.

Augmented Reality   *for Company's extrast power meetings*

The big selling point of AR is that it allows an expert sitting somewhere else to "augment" the reality you are looking at through a video screen on your phone, tablet, or laptop. They can explain what you need to do almost as if they were standing by your side. Here's how it works.

Your screen and the expert's screen show exactly the same thing—generally the scene you are looking at. The expert can then "augment" your reality—that is, the image on your screen—by placing computer graphics on it. These graphics appear as if they are really in the scene you are videoing with your phone or tablet. This makes communication much easier. Instead of talking you through it, they show you with arrows, circles, and the like. Instead of trying to describe which bolt you should loosen, button you should push, or sentence you should focus on, they show you. There is no need to "paint a word picture" of what needs to be done; the expert can paint a real picture. This clearly has many real-world applications, but it first got popular as a game.

You probably have already heard about AR although not under that name. You probably have heard of it as Pokémon Go. This video game became wildly and almost instantly popular when it launched in July 2016. It broke five records in the Guinness Book of World Records. It was

13. Stephane Kasriel, "This Is What Your Future Virtual-Reality Office Will Be Like," *FastCompany.com*, July 19, 2016.

downloaded 130 million times in its first month. The game, which runs on smartphones or tablets, overlays a fantasied version of your neighborhood on your screen. Not a fantasy neighborhood, your real neighborhood, be it Trafalgar Square, the Empire State Building, the Eiffel Tower, or Tokyo Station. The game uses GPS to know where you are.

When you get close to certain places, it "augments" the reality you see on your phone's video screen. For example, with your naked eyes, you see only a park bench in Central Park. But on your screen, you see a 3D, animated cartoon character jumping around on the bench. Your mission, should you choose to accept it, is to capture the Pokémon with a Pokéball. If that doesn't make sense, ask one of the hundreds of millions who have played the game.

The AR that is being used for work is much less sophisticated than Pokémon Go. Instead of a computer program sending 3D cartoon images onto a smartphone or tablet screen, companies are using AR to provide expert advice to workers in the field when, for example, field workers have to repair a piece of equipment they've never seen before. This is a new form of two-way communication that makes workers feel like they are working side by side even when they are far apart.

This is not science fiction and the technology isn't even very fancy. Most of today's applications use smartphone or tablet screens, but there are also specially made headsets that allow hands-free communication.[14] It is also being used for group meetings.

These new forms of communication make videoconferencing and video Skype look positively Neanderthal. They are going a long way toward taking the remote out of remote work. To date, most of the uses have been in situations where it is almost impossible to have workers side by side. And most applications have involved domestic remote work. For example, Dutch police are using AR to help first responders deal better with crime scenes they walk into as part of their job.

---

14. One that stands out—but is not really mainstream yet—is Microsoft's HoloLens. This is basically a laptop that you wear on your face like a pair of goggles, so you can see the real world with digital images overlaid.

DUTCH POLICE AND GAZA STRIP SURGERY

Firefighters, and paramedics are often the first ones to arrive at a crime scene. Usually they have more important things on their mind than preserving evidence. Even if they did have the time, they are unlikely to have the training needed to document crucial evidence, procure samples, or check whether perpetrators are still at the location. These experts need help from other experts, but it is impossible to send crime-scene experts along with every ambulance.

To get around this limitation, Dutch police are using AR. The first responder, say a paramedic, wears a camera and a smartphone that establishes two-way communication with a crime scene investigator located elsewhere. The investigator can point out objects that the paramedic should avoid touching as they may be critical to the subsequent investigation. This is not done by describing the object; it is done by electronically placing a circle over the object on the screen of the paramedic's smartphone.

The circle then instantly appears on the paramedic's screen and through the magic of image processing, the circle stays fixed on the indicated object even as the paramedic moves around or pans away from the object and then pans back to it. It is easy to see how this provides a reasonable substitute to working side by side, even when the two people are far apart. This is a game changer since it makes the two-way communication surer and faster.

As with all these new communication technologies, the result is not as good as having the expert physically standing next to the worker in the field. But getting expert advice is a whole lot cheaper and faster with AR. From the perspective of the expert, AR opens up many more opportunities for selling his or her particular expertise. With AR, an expert mechanic, for example, could provide advice to many different repairers without ever traveling.

Surgery is another area where AR is already being used. One example is the application Proximie which allows a surgeon in one place to help a surgeon in another place. The remote surgeon guides the operating surgeon with screen markings that point out things like tendons, arteries, nerves, or where to make the incision. Proximie, which has been in use since 2016, has been used by doctors in Beirut to assist surgeons operating in the Gaza Strip. And remote surgical assistance via AR is not only for war zones. The headwear known as Google Glass (which looks like a pair of glasses) has been used in cardiac procedures in ways that allow an expert in a specific procedure to provide real-time advice to the operating surgeon.

The other main new form of communication, virtual reality (VR), is a far more immersive experience—it completely hijacks your visual and audio channels filling them with a computer-generated reality. It can be a bit disorienting since you have no direct connection with where you are actually sitting.

Experimental Communication Technologies

There has been a lot of hype about VR. And it may be one more case of a technology that is being overhyped. But before dismissing it, it is worth watching some of the demos on YouTube and imagining how this technology would make it easier to work with faraway people. Or better yet, try out a VR headset yourself.

To date, the images are quite grainy, but the body language that comes through has amazing effects on how you perceive people. I tried a VR workplace system at an IHS Markit event in London in May 2017. It was a virtual trading platform (a workstation for people trading financial securities). The scientist who was demonstrating it talked me through the features while I was wearing the headset and when he was done he said: "Do you want to come out now?" And when I took the headset off, I had the very distinct impression of leaving one room and entering another. In this case, there was no one else in the virtual room with me, but it doesn't take a lot of imagination to see that I could have been having a virtual meeting with other people wearing similar headsets.

There are other forms of ACTFAR in testing stages. Many seem to be drawn directly from episodes of *Star Trek*. The next step in almost-being-there communication is "holographic telepresence." This projects real-time, 3D images of people (along with audio) in a way that makes it seem as if the remote person is right next to you. This is the stuff of science fiction, but it is not unimaginable.

In 2017, the French presidential candidate, Jean-Luc Melenchon, campaigned in Lyons and Marseille at the same time using a holographic projection. In 2014, the prime minister of India, Narendra Modi, also used holographic presence to be at far more campaign rallies than he could have done in person.

Microsoft's Holoportation—and other similar products by Cisco and Google—aim to mainstream this in coming years. Holoportation—a conscious play on the teleportation of *Star Trek* fame—is a form of virtual reality that makes people seem as if they are in the same room even when they are physically in distant places. Specifically, it projects a hologram video image of a person Who is in one room into another room. The people in the two rooms can interact with anyone who is in either room almost as if they were actually there.

The technology uses lots of cameras and high-powered processing to transform videos of people into realistic 3D models in real time. The system then transmits the models to the headsets of people in another room (it works best if the two rooms are perfect copies of each other). In early 2016, the system was enormously bulky but by the end of the year, Microsoft shrunk the gear down far enough to get into a minivan and reduced the bandwidth requirements by 97 percent, so it could work on standard, high-quality Wi-Fi networks.

The YouTube videos demonstrating Holoportation are remarkable, to say the least. If this Holoportation ever becomes mainstream, it would radically transform the meaning of telecommunication. It would make it much easier to interact with people across the world. Or, to put it differently, your company could hire foreign professionals willing to work for small money, or you could export your expertise across the world without leaving your desk.

A different technical approach projects a standard hologram into a remote room. ARHT Media, for example, has a service that projects speakers virtually via what they call "HumaGrams"—which are like telegrams for humans. The technology, in use since 2015, allows speakers to be virtually present in front of an audience far away.

AR and VR are especially helpful in situations where two or more workers have to interact with something physical. But a great deal of work in offices depends upon regular meetings. As it turns out, digitech has created a marvelous substitute to actually being physically in the same room as other workers—it is called a telepresence robot. One company that is using it today is the online media site, Wired.com.

TELEPRESENCE ROBOTS

Emily Dreyfus writes for the San Francisco company Wired.com but lives in Boston. She used to participate remotely in staff meetings and bilaterals with her editor in the usual twentieth-century way—by phone, messages, and video conferences. But this wasn't good enough for the spontaneous, creativity-enhancing brainstorming sessions that Wired was hoping for.

Being a northern Californian sort of company, they decided to throw some digital technology at the problem. The tech took the form of a "telepresence robot" made by Double Robotics. The movements of the telepresence robot, which you can think of as Skype on wheels, are controlled by the writer in Boston, so the robot (in San Francisco) can wander around the office, attend meetings, hold one-on-one meetings, and so on. Picture the robot as a normal sized iPad on a stick with the stick attached to a Segway. It has a forward-looking camera, a microphone, and speakers. Dreyfus, whose face fills the iPad screen, can drive it around the San Francisco office at strolling speed.

At first, the whole thing seemed strange to Dreyfus—as new technologies usually do. But she soon fell in love with it. She even gave the robot a name, "EmBot." Dreyfus found that other writers and her editor responded much better to her when she was "in" EmBot than they did when she was on the phone. During staff meetings, she felt connected to the others in a way that was impossible before. She would turn to "face" whoever was speaking. "The crazy thing about being a human 3,000 miles away from your telepresence robot is that the divide instantly dissolves when you activate. As soon as I call into EmBot, I am her, and she is me. My head is her iPad. When she fell, I felt disoriented in Boston. When a piece of her came off in the impact, I felt broken."[15]

And the feeling was reciprocal. The robot gave her a physicality that the other workers instinctively treated as a real person who was really there. Or almost. There was that case of inappropriate robot touching.

On one of the Embot's first days at work, an office joker moved behind her screen while she was chatting, picked up the robot, and shook it. This "inappropriate robot touching" made Dreyfus feel violated, powerless. There are now rules at Wired: no touching robots without the telecommuter's permission. The rules, however, only apply when EmBot is activated. If Dreyfus's face is not "on," it is considered no more alive than a broomstick. Dreyfus intentionally goes offline if someone has to carry the broomstick somewhere, like the charging station.

The deep reason EmBot is so effective has to do with evolutionary psychology.

The Mind Bugs behind Telepresence Robots

Things that move have meaning—or at least that is our lizard brain's first instinct according to social psychologists. This was powerfully illustrated by one of the most famous experiments in psychology. People watched a one-minute film of three shapes—one large and one small triangle and a circle—that moved in and around a big rectangle that opened and closed. These shapes did not look anything like people.[16] The researchers, Fritz Heider and Mary-Ann Simmel, then asked people to describe what they had seen.

15. See Emily Dreyfuss, "My Life As A Robot," *Wired.com,* September 8, 2015.

16. F. Heider and M. Simmel, "An Experimental Study of Apparent Behavior," *American Journal of Psychology* 57, no. 2 (1944): 243.

Without any prompting, most participants assumed that the geometric shapes represented humans, and they made sense of the movement by projecting human motives onto the colored shapes. Try it yourself; it is easy to find the Heider-Simmel video online. See if you interpret the clip as a love story of the type you might expect in one of those old movie Westerns. Many of the participants in the experiment interpreted the circle as a woman who was in love with the little triangle, taking the big triangle to be a larger man who tries to steal her love.

Social psychologists call this very human reaction "attribution." People attribute motives and meaning to physical movement of any object—especially when the thing is physically present. It is why some people name their car, but few name their iPhone even though they sit in their car and talk to their phone.

Believe it or not, the Heider-Simmel experiment tells us something about why telepresence robots are catching on fast. Many hospitals and some companies use telepresence robots already, and their use is growing rapidly since the impact on team interactions is palpable. The sense of being face-to-face is much stronger when the face moves, so to speak. In particular, doctors find that their words carry more authority with patients when they are talking via a telepresence robot instead of normal video Skype, or over the phone.

While telepresence robots are useful for many interactions, a static form of telepresence technology is transforming the ease of holding meetings over long distances.

Fixed Telepresence Systems

Telepresence systems—a static version of EmBot, if you will—are already widely used by big banks, consultancies, law firms, and governments. The high-end systems are still expensive. Telepresence rooms can run into the hundreds of thousands of dollars. But as the digital laws advance and construction moves into mass production, telepresence will get much cheaper and more mobile. It will accelerate the trend toward telemigration.

Think of standard telepresence as extremely good Skype—but so much better that it becomes a new experience. Telepresence makes it almost seem like people are in the same place even when they are not. I used it in spring 2017 to present my book, *The Great Convergence*, to the Norwegian sovereign wealth fund, Norges Bank Investment Management (NBIM).

I was in London with a couple of analysts and connected via telepresence with another group of NBIM economists located in New York City and with yet a third group in Oslo. At first it seemed like nothing more than Skype with a really good screen. But that soon changed. I could see that the remote participants were reacting to what I was saying and to my hand and facial gestures just as if they were in the same room. And they, I assume, had the same impression. The sense of personal connection jumped up a level. It was almost as if we were all in a single room.

The key is how telepresence plays on our brains' social "hardwiring." Everyone's brain is a like a high-powered computer when it comes to social interactions. Deciding whether to believe and trust others was a key evolutionary skill. As Steve McNelley and Jeff Machtig—founders of an edgy telepresence start up, DVE—put it, humans "have mastered the gathering and processing of nonverbal communication cues. It is second nature to us, and it is foundational to who we are and how we see others. It is an essential part of our humanity."[17] Thanks to life-size images on the screen, excellent image resolution, and superior sound quality, telepresence transmits much more of this nonverbal communication than does, say, Skype or Facetime.

Telecommunication is only one element of the technology that is used to knit together remote teams. Recent advances in so-called collaborative platforms are also making it much easier for workers to telemigrate.

17. See Steve McNelley and Jeff Machtig, "What is Telepresence?," undated article on *DVETelepresence.com*; visited June 25, 2018.

How Collaborative Software Facilitates Remote Work

Email is the granddaddy of all collaborative software. It—and the ability to share editable files (documents, spreadsheet, presentations, photos, videos, and the like)—changed the world and made it radically easier to work with faraway people. While email is fantastic (and irreplaceable since everyone uses it), it is deeply flawed as a means of coordinating teams. Its basic design choices were made when Bill Clinton and John Major were in power. Some of these choices are not optimal for today's world of work. Just ask anyone under twenty-five what they think of email, and you'll get my point.

The new collaborative platforms that firms are embracing—things like Business Skype, Slack, Trello, Basecamp, and more—are not perfect, but they reflect fresh, thorough, and highly intelligent thinking on how best to organize communication among team members. These new collaborative platforms are designed to facilitate all manner of team communication—everything from text chats, emails, and discussion groups to phone calls, Facebook posts, and multiperson video calls with screen sharing. Slack is one of the most popular and fastest growing platforms, but it has plenty of rivals including Facebook's Workplace, Microsoft Yammer, Google Hangouts, Microsoft's Teams, and a number of start-ups like HipChat, Podi, Igloo, GitHub, and Box.

Also related is another set of new organizational tools that are not so new: project management software. Some of these have been around for years, but many (Wrike, Microsoft Projects, Basecamp, Workfront, etc.) are now designed to work with geographically dispersed teams. There are also tools, like Mural, that assist remote collaborative design efforts and brainstorming. The tools in this "space" are developing rapidly, but they have already radically lowered the difficulty of weaving remote workers into projects.

When it comes to bringing foreign competition directly into American and European offices, all this new technology is important. But at least as important is that fact that we and our companies are rearranging things to make telecommuting easier. To date, most of this telecommuting takes place domestically but it doesn't take a lot of imagination to see that domestic telecommuting can easily become international telecommuting.

Domestic remote work is the thin edge of wedge that is opening the service sector to telemigration. And it is astounding how many jobs are already being done remotely.

DOMESTIC REMOTE WORK PAVING THE ROAD FOR TELEMIGRANTS

David Kittle is an industrial designer who feels strongly about his creations. Products should be functional and aesthetically interesting—an approach that has helped him develop winning designs for just about everything from rugged electric lanterns and plastic playground equipment to motorcycle cup holders and roller-coaster seats. "It's pretty cool when someone hands you a dream and you are able to hand it back over to them in real life. There is a lot of joy in that," he notes.

Amazingly, David does all this from home. You can hire him online for $150 an hour.[18] David is most definitely not alone. Using remote workers to get jobs done makes sense financially and personally for people like Kittle and the US companies that hire him. But the trend has unintended consequences for all domestic service-sector workers. It is the first step towards direct international competition among freelancers—and freelancing is a trend that is big and growing fast.

Government statistics tend to misclassify remote workers, so surveys are the best way to measure the trends. A recent Gallup poll asked questions about all types of remote work—not just the full-time freelancing that Kittle is doing. It found that 43 percent of US workers telecommuted sometime during 2016—four times more than in 1995—and they are doing it more days per week. About a fifth of the telecommuters work remotely

18. Melanie Feltham, "Spotlight on David Kittle, Top Rated Freelance Product Designer," *Upwork* (blog), July 19, 2017.

full time. Under the Obama administration, almost one in three federal employees worked from home at some point during 2016.

A 2016 survey by an organization that supports US freelancers estimated that fifty-five million Americans—that's 35 percent of total workforce—were freelancing. That is a couple of million more than the estimate from the 2014 version of the survey. As you might imagine, younger people are more likely to be freelancing. In the eighteen- to twenty-four-year-old group, almost half are freelancing at least part time or full time. Indeed, many of the millennials (workers under 35) in the survey have never had a traditional job; they have spent their entire working careers as freelancers. Among baby boomers, it is rather less common.

Another factor that is accelerating the trend toward remote work is the way US and European companies are reorganizing themselves to accommodate telecommuting workers.

The Dissolving Office

Traditional offices had all the workers and bosses in the same building. Everyone showed up at the same time; coffee breaks and lunchtimes were synchronized. This helped the bosses establish hierarchies, it helped teams work together, and it helped colleagues to trust each other. The phrase "I have to get to work" meant you were going somewhere, not just doing something. Digital technology changed this.

Technology has allowed companies to adapt faster to changing demands. But the ability to adapt quickly has, in turn, spurred the demand for more rapid responses. Customers can switch suppliers and products more quickly. The services in demand are shifting more rapidly. New competitors are springing up in ways that were previously impossible. This onslaught of competition has undermined the old static hierarchies, fixed desks, and demands for physical presence, and fixed hours. Routine processes are being replaced by "agile," project-oriented corporate structures with flatter management profiles and cross-department teams (sometimes called a "matrix" structure).

To adapt to rapidly changing challenges and opportunities, firms are moving away from traditional employer–employee relationships. Increasing reliance on remote workers (especially those who are not traditional, full-time employees) is providing today's service-sector companies with essential elements of flexibility.

"To keep pace with constant change in the digital era," noted the Accenture *Technology Vision 2017* report: "The future of work has already arrived, and digital leaders are fundamentally reinventing their workforces. . . . The resulting on-demand enterprise will be key to the rapid innovation and organizational changes that companies need to transform themselves into truly digital businesses." There is a lot of business-school jargon packed into those sentences, but you should latch on to the basic point: steady jobs won't be so steady anymore.

If this were the cable entertainment industry, we would call this the "pay-per-view model of work." Companies will browse online for the workers they need and pay them per project—as the need arises. The number of employees can grow fast to seize opportunities, but can also shrink fast when exiting losing adventures. Remote work is a key element of this vision. It also means shifting work organization to cloud-based platforms that allow people to work anywhere anytime. Much of this is already a reality.[19]

One really radical thinker—and one who was years ahead of the curve—is Michael Malone. His 2009 book, *The Future Arrived Yesterday*, projected a world where the "Protean Corporation" has only a small set of core people on long-term contracts with all the rest done by outsourced providers. The US company Snapchat is not far from this. It was worth sixteen billion dollars in 2017 but had only 330 employees. To understand just how different this is, consider the same figures for a traditional

---

19. The massive multinational, GE, is an example. It is moving away from location-based hierarchical decision making to something that looks more like a start-up organization project-by-project. GE even has a snappy double-meaning tag for it. It is called "FastWorks." This, the company claims, allowed them to build a diesel engine for ships that meets new environmental regulations a couple of years ahead of their competition.

corporation. General Motors is worth about fifty billion dollars and employs 110,000 workers worldwide.

The buzzword phrase that Accenture has developed to describe this future of employer–employee relationship is telling. They call it the "liquid workforce." For now, much of the "liquid labor" is hired domestically, but there is plenty of liquid labor abroad eager to work for a fraction of US and European wages. This sort of corporate reorganization, in short, is opening another lane of the cyber highway that will bring American and European service and professional workers into direct competition with telemigrants.

All these things are creating snowball effects. As more workers work remotely, companies adjust their work practices and team structures to make this easier, and as it gets easier, more workers do it. This in turn has stimulated digital innovations that facilitate remote work. The snowball has created a hundred-billion-dollar business sector for the technology and services that grease the wheels of remote work.

There is, in a sense, the equivalent of a "reverse industrial revolution" going on in offices. In the first phase of industrialization, textile work moved from cottages to large mills. Now office work is moving from large offices to the twenty-first-century equivalent of cottages.

A key question is, which jobs will be displaced by this white-collar globalization?

WHICH JOBS WILL BE DISPLACED BY TELEMIGRANTS?

The easy route to answering this question is to just look at all the jobs in which people are working remotely today—usually from within the same city, or at least the same country. Just look around you and see which types of jobs lend themselves to remote work and you'll get an idea of where competition from foreign freelancers is likely to hit soonest and hardest. The harder route is to think about the tasks involved in each occupation and then think about which of those could be done by a talented foreigner sitting abroad.

"You have to be there" is a key part of the job description for occupations like childcare workers, farmers, and surveyors. These sorts of jobs cannot be done by workers abroad since the very nature of the job requires a physical presence. But which jobs are these? Thanks to the research of Princeton professor Alan Blinder, we can be more specific.

Alan Blinder is an intellectual who cares. He is the epitome of a policy-relevant economist using his specialized knowledge to make the world a better place. The title of his 1988 book, *Hard Heads, Soft Hearts: Tough-Minded Economics for a Just Society*, says it all. And he put both his hard head and soft heart to work in the 1990s serving as vice chair of the US central bank, and a member of President Clinton's Council of Economic Advisors.

In the 2000s, Blinder became passionately concerned by the possibility that advancing information technology—what today we call digital technology—could lead to the loss of US jobs due to offshoring. What he had in mind is reverse telemigration. Instead of foreign workers working virtually in our offices, he was concerned that "our" work would be sent to foreign offices. And in many areas like call centers, and back-office processing, that is exactly what happened.

As part of his effort to raise the alarm, he developed a ranking of how "offshorable" each US occupations was. His ranking was based on two criteria. If the job had to be done at a specific location in America, then it could not be displaced by foreign competition. If the job could be done remotely, Blinder assigned a numerical value to how easily the output of the work could be transmitted with little or no deterioration of quality.

Using these criteria, he estimated that about half of all management, business, and financial jobs could be done from abroad. The share was about 30 percent for many professional, and office and administrative jobs. In terms of sectors of the economy with the most offshorable jobs, Blinder lists professional, scientific, and technical sectors as having almost 60 percent of the jobs open to international wage competition. In finance, insurance, and the media, half of the jobs are vulnerable. According to popularizations of his study (which dropped his careful hedging), anything that could be sent down a wire would eventually be offshored. And remember, that was in a different era of technology. It was before digitech removed much of the "remote" from remote work.

Subsequent studies tweaked these estimates, but the new numbers remain in the range of one in three US jobs. That is a scary number. If even half the workers holding these jobs today came into direct competition with foreigners in a few years, there would surely be a mighty upheaval—and a cry for shelter from the shocks.

White-collar globalization is an amazing thing. It will change our lives. But it is only half of the dynamic duo that is driving the Globotics Transformation. The other half is white-collar automation.

6

Automation and the Globotics Transformation

James Yoon is a prosperous Californian. He has a good job working as a lawyer specializing in patent disputes. There is lots of work since the tech giants are forever squabbling over who invented what first. Today, he charges them $1,100 an hour.[1] That's way up from the $400 he charged in 1999, and his price is up not just because he is older and wiser. The nature of his work has been transformed by digital technology, specifically by AI-trained computer programs.

At the end of the twentieth century, a big patent dispute would involve three partners (the head honchos at law firms), five associates (the deputy honchos), and four paralegals (the assistants). That's eight lawyers and half as many highly skilled assistants. Today, Yoon would be the only partner and he'd use only two associates and a single paralegal. The legal talent was cut to a quarter of its previous level.

1. Steve Lohr, "A.I. Is Doing Legal Work. But It Won't Replace Lawyers, Yet," *New York Times*, March 19, 2017.

How does Yoon cope with the radically lower headcount? The answer is certainly not that the law got simpler or the paper trails shorter. The answer is that white-collar robots have taken over some tasks—especially those that can be thought of as "knowledge assembly line" functions. Robo-lawyers are good at things like searching through documents and emails and flagging which ones will be relevant.

Yoon uses two robo-lawyer programs (Lex Machina and Ravel Law) to help him plow through information that suggests the type of legal strategy he should employ. These bits of software can get their "mind" around huge piles of court decisions and the documents filed on similar cases by the judges and opposing lawyers. Robo-lawyers cannot do it all, but some of the legal talent is being displaced. Indeed, displacing human lawyers is one of the main attractions of using robo-lawyers. This is one reason that Yoon is thriving.

Robo-lawyers are just one example of how AI-trained white-collar robots are driving the Globotics Transformation.

MEET WHITE-COLLAR AUTOMATION

The sophisticated computer systems and machine learning algorithms that are behind Lex Machina and the like are very expensive and require PhD-level computer scientists to get them up and running. If these so-phisticated AI platforms were restaurants, they'd have a Michelin star or two. This puts them out of the reach of the companies for which most people work, namely small- and medium-sized firms. There is, however, a "fast-food" version of white-collar robots. It's called "robotic process au-tomation" (RPA) software; Poppy, who we met in Chapter 4, is a good example.

RPA is probably not what comes to mind when people speak of the "robot apocalypse," but RPA will be a key part of the Globotics Transformation. It's worth a closer look. RPAs are automating white-collar jobs in a very direct way.

The Low-End Competition: RPA

"They mimic a human. They do exactly what a human does. If you watch one of these things working it looks a bit mad. You see it typing. Screens pop-up, you see it cutting and pasting," explains Jason Kingdon, chairman of one of the leading RPA companies, Blue Prism. They are designed to be "an automated person who knows how to do a task in much the same way that a colleague would."[2]

This is why Blue Prism describes their RPA programs as "robots" instead of software. They are synthetic workers, in essence. This type of AI aims to cut jobs for people involved in the back-office processes commonly found in finance, accounting, supply chain management, customer service, and human resources. RPA robots are remarkably simple to implement.

"They're easy to use and have a relatively low cost," says Frances Karamouzis, who is research vice president of the IT research firm Gartner.[3] Adoption of RPA is booming. One consultant company, Transparency Market Research, expects RPA implementation to grow at 60 percent per year worldwide through 2020. Another market research organization puts the figure at 50 percent per year. That is explosive growth. And the growth is coming for good reasons.

First, RPA robots are much cheaper than humans. The Institute for Robotic Process Automation estimates that an RPA software robot costs a fifth of local workers, and a third of offshore back-office workers located in, say, India. Second, the work is more consistent, and it leaves a digital trail that makes reporting for regulatory compliance reasons faster and surer. Third, the processes can scale up and down rapidly to deal with, for example, seasonal fluctuations in the paperwork flow; there is no need to hire and train temporary workers when you can just run the software a bit harder.

2. Hal Hodson, "AI Interns: Software Already Taking Jobs From Humans," *NewScientist.com*, March 31, 2015.

3. Bob Violino, "Why Robotic Process Automation Adoption Is on the Rise," *ZDNet.com*, November 18, 2016.

In some sense, RPA is the "wave of today" when it comes to globotics automation. The "wave of tomorrow" refers to the more sophisticated systems—the Cortanas and DeepMinds of this world. These can handle a much wider range of workplace tasks. This makes them a much deeper threat to existing human jobs, but it also makes them harder to implement and thus slower to phase in.

High-End White-Collar Robots

Amelia, the white-collar robot we met in Chapter 1, is not just an amazingly productive service-sector worker, she is simply amazing. Research had shown that customer satisfaction with phone-in helplines is directly tied to empathy shown by the agent handling the call, so Amelia's maker added a psychological module to her algorithm. She is thus aware of the emotional state of the person with whom she is speaking, and she adapts her responses, facial expressions, and gestures to better communicate.

In her most advanced version, where customers are using smartphones or laptops with cameras, Amelia uses facial recognition to begin new conversations. The customers are not treated as strangers but as acquaintances; Amelia begins new conversations with the full knowledge of all a customer's previous contact history.

When Amelia can't handle something, she passes on all relevant information to her human colleagues so they can continue. But Amelia is curious. The software hangs on the line listening to what the humans are talking about—especially the resolution of the problem. She then adds these new tricks to her knowledge management system. Once her learnings are approved by her human supervisor, she can answer similar queries herself in the future.

Just in case you think Amelia is a flash in the pan (like many AI wonders have been in the past), it's worth noting that Amelia is used by over twenty of the world's leading banks, insurers, telecom providers, media companies, and healthcare firms. And she has rivals. Since 2016 or so, many companies have been introducing Amelia-like software.

Bank of America rolled out Erica in the summer of 2018. She offers one-to-one services that are usually reserved for bank customers with bulging balances. (Or actually, the high flyers will still get one-on-one services; the masses get one-on-Erica services.) Erica address Bank of America customers by first name on their smartphone or ATM machines. She can, for example, let you know when your checking account has dipped into the red. But she knows much more about you than just your balance. She uses AI to make helpful suggestions: "Based on your typical monthly spending, you have an additional $150 you can be putting towards your cash rewards Visa. This can save you up to $300 per year."[4]

JPMorgan's white-collar robot is called Contract Intelligence (COIN), and Capital One has Eno. IBM is selling many Amelia-like virtual assistance under the brand Watson; Salesforce offers Einstein; SAP has HANA; Infor has Coleman; and Infosys has Nia. The public sector is getting in on the act too. The Australian government's cognitive assistant is called Nadia. She helps citizens get information services for the disabled.

Microsoft has Cortana, and Amazon has Alexa—the white-collar robot that "lives" in Echo (Amazon's home AI system). Apple's AI robot is the famous Siri, although she has not yet been deployed in workplace automation. Google has long used AI inside the company; the whole search engine, for example, can be thought of as a white-collar robot that has no particular name. If you want to talk to the nameless search-bot, you just say: "Hey Google."

The "nobility" of AI systems like Amelia, Watson, and Erica—together with the "squires" of AI systems like RPA—will displace many service-sector jobs. The big question is—which jobs? To answer the question, it is necessary to change gears a bit, since white-collar robots are not really taking whole occupations; they are taking over some of the activities that make up part of many occupations. This is a critical insight into the future of work.

4. Harriet Taylor, "Bank of America Launches AI Chatbot Erica — Here's What It Does," *MONEY 20/20, CNBC.com*, October 24, 2016.

ROBOTS WILL ELIMINATE MANY JOBS BUT FEW OCCUPATIONS

Think of your occupation as an imaginary to-do list, a list of "chores" or tasks, a catalogue of the things you have to do to get the job done. Keep in mind that this is not a static list—it is continuously evolving.

In recent years, great technical advances like laptops and smartphones—teamed with much better software and great websites—have substantially lengthened our to-do lists. Now, we are all our own typists, file clerks, travel agents, receptionists, and so on. In my father's day, a separate human performed each of those tasks. Now they are chores that I, and many other professionals, have to do ourselves. But things that can be bundled can also be unbundled.

Robots can take over some of your tasks, but not all. This means that you'll be more productive—and that may mean there will need to be fewer people like you doing the job—but robots won't eliminate your occupation. After all, most occupations involve at least some tasks that require a real person. Yet white-collar robots will reduce the headcount. It is just a matter of arithmetic.

Suppose an IT helpdesk at a bank gets a hundred requests per day. To handle these, the bank needed, say, ten workers. When online chatbots take over some of the chores that had been on the to-do list of each of the ten workers, the hundred requests can be handled by fewer than ten people. If the pile of work doesn't expand sufficiently, the result will be job loss.

There is really nothing new about this jobs-not-occupation point—it is what automation has always done. Tractors, for example, automated some farm chores, but they did not eliminate farming as an occupation. We just needed fewer farmers. This is what we'll see all across the service sector in coming years. And it is a critical point in preparing for the upheaval; white-collar robots will eliminate many jobs but few occupations.

From this tasks-not-occupation perspective, the next step in thinking about the which-jobs-will-go question is to work out what white-collar robots are really good at already. This is not an easy task.

White-Collar Robots' Work-Relevant Skills

There are over eight hundred different occupations, according to US government statistics—everything from animal trainers and CEOs to rock
splitters and roof bolters. And each of these jobs involves many skills. We
need to simplify to clarify. Here is where management consultants come
in handy.

The business and economic experts at the McKinsey Global Institute
have very usefully classified all workplace skills into just eighteen types.
To simplify I have classified McKinsey's eighteen skills into four broad
categories: communication, thinking, social, and physical skills. The
McKinsey experts looked at AI abilities in 2015 and assigned a grade to
the technology's ability to perform each of the eighteen skills. Since this is
necessarily a rough-and-ready judgment call, they handed out only three
types of grades. AI was judged as being able to perform: 1) at a level below
that of the average person ("below"); 2) at the level of an average person
("equal to"); or 3) at the level of a highly skilled person, i.e., someone
in the top 20 percent of the skill range ("above"). What they found is
fascinating—and a bit disturbing.

COMMUNICATION SKILLS

In most jobs, workers have to be able to understand what others are saying
to them. The McKinsey term for this is "natural language understanding."
White-collar robots are good at this, as most people will already know if
they have talked to Siri, Alexis, Cortana, or others of their kind. But it is
important to keep in mind that these software robots are not listening in
the human sense of fully comprehending what the words mean. Speech
is just particular patterns of sound waves. The computer digitizes these
and then uses its machine-learning-trained statistical model to guess at
which words are being spoken. It then interprets the words as speech
by looking for word patterns in terms of phrases and phrases in terms
of meaning. When the training data sets get big enough and computers
powerful enough, white-collar robots may be able to understand everything we say, but so far there are still many misunderstandings. That's why

McKinsey graded these AI's language-understanding skills as below the average human (see Table 6.1).

When it comes to speaking ("natural language generation"), AI is much better so AI's capability is graded as equal to an average human. The reason, as we saw with the Siri-learning-Shanghainese example in Chapter 4, is that speaking is much simpler for machines to master. The next communication skill is more specialized—crafting nonverbal outputs.

There are more ways to communicate than speaking and writing. Millions of jobs require people to produce videos, slideshows, presentations, or music. These are really just alternative forms of communication, and they are things AI programs are increasingly doing for us. One example of this is the slideshows that Facebook's bots suggest to users on occasion. The AI inside recent iPhones does a similar thing with photos. When it comes to

Table 6.1 Capabilities of AI in Communication Skills

Communication Skill	Description	AI Skill vs. Human Average
Natural language understanding	Comprehend language, including nuanced human interaction	Below
Natural language generation	Deliver messages in natural language, including nuanced human interaction and some quasi language (e.g., gestures)	Equal to
Craft non-verbal outputs	Deliver outputs/visualizations across a variety of mediums other than natural language	Equal to
Sensory perception	Autonomously infer and integrate complex external perception using sensors	Equal to

SOURCE: Author's elaboration based on data published by McKinsey Global Institute in "A Future That Works: Automation, Employment, and Productivity," January 2017, Exhibit 16.

this skill, "craft non-verbal outputs," the McKinsey experts rank AI tech-nology as getting a grade of "equal to."

The last communication skill is "sensory perception." This refers to skills that use various sensory inputs in working out what is going on. It is, in essence, "communication" with the physical objects around us. This is critical in many jobs. In most jobs, we have to recognize objects and patterns by seeing, hearing, and touching. Self-driving cars have to recognize objects on the road and distinguish between a dog sitting in the road and a speedbump. A robot that lifts and puts an elderly person in a wheel chair has to feel when they have the person in their robot arms. On these skills, AI gets a passing grade—their performance is judged on par with average human capabilities.

Taken together, these four communication skills are what you might think of as the "gateway" skills—the capacities that open the gate to us using white-collar robots more widely at work. Yet, their communication skills are not why white-collar robots will be so disruptive to service jobs. The really disruptive thing is their inhumanely good ability to recognize patterns based on unimaginable amounts of experience (data).

THINKING SKILLS

Thinking skills are part of basically every job in the service sector that hasn't already been replaced by a machine. But there are many types of thinking. At one end of the spectrum is "creativity," at the other end is hardcore logical reasoning. In between, the McKinsey experts singled out "identifying new patterns," "optimizing and planning," "searching and retrieving information," and "recognizing known patterns" (see Table 6.2).

The level of thinking skills that AI has, according to McKinsey, is below the human average for creativity, identifying new patterns, and logical reasoning and problem solving, but above human average in planning, searching and retrieving information, and recognizing known patterns.

Keep in mind that this comparison of the talents of humans and white-collar robots is unidimensional. These robot-talents are based on what AI experts call "narrow" intelligence. The algorithms behind the skills are the digital equivalent of a one-trick pony. Humans, by contrast, have

Table 6.2 CAPABILITIES OF AI IN THINKING SKILLS

Thinking Skill	Description	AI Skill vs. Human Average
Creativity	Create diverse and novel ideas, or novel combinations of ideas	Below
Identify new patterns	Create and recognize new patterns/ categories (e.g., hypothesized categories)	Below
Optimization and planning	Optimize and plan for objective outcomes across various constraints	Above
Search and retrieve information	Search and retrieve information from a large scale of sources (breadth, depth, and degree of integration)	Above
Recognizing known patterns	Recognize simple/complex known patterns and categories other than sensory perception	Above
Logical reasoning/ problem solving	Solve problems in an organized way using contextual information and increasingly complex input variables other than optimization and planning	Below

SOURCE: Author's elaboration based on data published by McKinsey Global Institute in "A Future That Works: Automation, Employment, and Productivity," January 2017, Exhibit 16.

"general" intelligence, which means we can think abstractly. We can plan for things that might happen and solve problems at a general level without nailing down all the details. Humans can innovate and develop thoughts and notions that are not based directly on past experience.

Computer algorithms that are trained by machine-learning techniques can't really "think" in the human meaning of the word—or even in the dog or pony meaning of the word. The AI is just taking in data and guessing at what the data corresponds to. It can do this "taking in" and "comparing" incredibly fast, but it can only recognize things it has seen in its training data set. This limitation can be illustrated with one of the edgy attempts to go beyond standard machine learning techniques—a form of machine learning called "unstructured learning." This is an approach where the computer identifies patterns on its own.

In one famous example of unstructured learning, Google set a computer system, Google Brain, loose on millions of clips from YouTube videos to see what patterns it would find on its own. In a feat that amazed the AI world, it did find a pattern and, given that it was looking at YouTube videos, it's not surprising that the pattern was a cat. Of course, the computer didn't know it was a cat—humans had to tell it that—but it recognized that all the images corresponded to the same object.

This form of machine learning may be important in the future, but for now it is problematic. One of the other things Brain identified as a "thing" looked like a combination of an ottoman and a goat.[5] No one really knows what it was thinking. For now the main applications use structured learning which requires a training dataset where the issue is clear ("Is this a face?") and the outcome is clear (yes or no).

This sort of limitation is why robots function poorly when there is little data to train the algorithm. For example, it is hard to generate a dataset for being creative, since the whole idea of creativity is to be somewhat unique, or unusual. Likewise, software robots aren't very good when the nature of the problem and the nature of the solution are just intrinsically vague. That's the case when identifying new patterns: the whole idea is that the pattern is new, so there cannot be a big dataset by definition. For example, a human Go master could presumable do fairly well on a slightly different-sized board, but AI couldn't. At a 2017 conference, the AlphaGo Master team admitted that the AI-software would be useless if the game was played on an even slightly altered board—say one that was

---

5. Gideon Lewis-Kraus, "The Great A.I. Awakening," *New York Times Magazine*, December 4, 2016.

Table 6.3 CAPABILITIES OF AI IN SOCIAL SKILLS

Social Skill	Description	AI Skill vs. Human Average
Social and emotional reasoning	Accurately draw conclusions about social and emotional state, and determine appropriate response/action	Below
Coordination with many people	Interact with others, including humans, to coordinate group activity	Below
Act in emotionally appropriate ways	Produce emotionally appropriate output (e.g., speech, body language)	Below
Social and emotional sensing	Identify social and emotional states	Below

SOURCE: Author's elaboration based on data published by McKinsey Global Institute in "A Future That Works: Automation, Employment, and Productivity," January 2017, Exhibit 16.

twenty-nine-by-twenty-nine squares instead of the standard nineteen by nineteen.[6]

The next set of work-relevant skills are social skills.

SOCIAL SKILLS

Many people are "socially tone deaf," and you probably have to work with some of them. They seem unable or unwilling to pick up on the little clues that someone is feeling down, overwhelmed, or elated about something and wants to share. White-collar robots are like that on the whole (Table 6.3.

These social skills are critical in occupations that involve a lot of interactions with people including coordinating with many people, and are important in work environments that require team work or management.

6. Ron Miller, "Artificial Intelligence Is Not as Smart as You (or Elon Musk)," *TechCrunch.com*, July 25, 2017.

The McKinsey experts graded AI-trained algorithms as having capabilities that are below that of the average person in all four of the social skills. This includes "social and emotional reasoning," "coordinating with many people," "acting in emotionally appropriate ways," and "social and emotional sensing."

It should be noted that improving the social skills of AI is an active area of research, so the McKinsey estimates may be a bit behind the times. The research is focusing on reading the social and nonverbal clues sent by individuals rather than social group dynamics. For instance, Disney is using machine learning to judge the reactions of movie watchers, specifically whether people laugh at the "right" time. To gather the training data, Disney's research team showed nine different movies a total of 150 times in a four-hundred-seat room that was equipped with cameras that monitored people's facial expressions. Disney gathered sixteen million face images.[7] The algorithm trained on this data was able to predict which expression a particular audience member was likely to make at various points in the movie after following that person's face for just a few minutes.

PHYSICAL SKILLS

Physical skills are important in a wide range of service-sector and professional jobs. Some of the physical skills involve moving things a long way ("gross motor skills") or over only very short distances ("fine motor skills/ dexterity"). Another set entails "mobility across unknown terrain," and "navigation." (Table 6.4)

Not surprisingly, industrial robots—what might be called "steel-collar robots" to contrast them with white-collar robots—are above average when it comes to most physical skills. They are, after all, machines. One area where they are not as good as the average person is in mobility in places they are not familiar with. Moving around an Amazon warehouse, for instance, poses no issues for AI-trained robots, but crossing rugged or unusual terrain is a skill where AI displays below-human capacities.

7. Disney Research, "Neural Nets Model Audience Reactions to Movies," *Phys.org,* July 21, 2017.

Table 6.4 CAPABILITIES OF AI IN PHYSICAL SKILLS

Physical Skill	Description	AI Skill vs Human Average
Mobility across unknown terrain	Move within and across various environments and terrain	Below
Fine motor skills/ dexterity	Manipulate objects with dexterity and sensitivity	Equal to
Navigation	Autonomously navigate in various environments	Above
Gross motor skills	Move objects with multidimensional motor skills	Above

SOURCE: Author's elaboration based on data published by McKinsey Global Institute in "A Future That Works: Automation, Employment, and Productivity," January 2017, Exhibit 16.

Having been properly introduced to AI software robots and having learned about what they are capable of, we now get to *the* question about white-collar automation. How many jobs will go? In fact, a number of researchers have developed estimates of how many jobs will be displaced. Think of these estimates as dogs walking on their hind legs: the interest lies not in that it is done so well, but rather that it is done at all. And I mean that with the greatest respect. Thinking hard about the future is not a mission for the faint-hearted, but it is something that society clearly needs.

HOW MANY JOBS WILL AI DISPLACE?

Many studies have tried to estimate the total impact of recent, AI-linked automation on jobs. These are essential reading but far from infallible. They are, after all, talking about the future, which means they are making it up—making it up using sophisticated methods and the best available data, but still, they are guessing.

Before getting down to details, here is the main takeaway. Over the next few years, the number of jobs displaced by white-collar robots will be somewhere between big and enormous. "Big" means one in every ten jobs is automated; "enormous" dials that up to six out of ten.

The granddaddy of these studies was done way back in 2013 by two Oxford professors, Carl Frey and Michael Osborne. They first got a list of all the chores involved in US jobs from a big US database maintained by the US government. Then they went through these and pegged the ones that they thought were automatable. They did this by starting with a list of tasks that were automatable and then calling out the occupations which depended on many automatable tasks. Half of all US jobs, they estimated, were at risk—yes, half (or 47 percent to be precise). The latest update of this approach—done by McKinsey based on the information reviewed above—raises this to 60 percent (due in part to the fact that white-collar robots have gotten so much better).[8] These rather startling numbers refer to jobs that could be automated. But how many actually will be?

A recent study by the consulting firm, Forrester, suggest that 16 percent of all US jobs will be displaced by automation in the next ten years.[9] That is one out of every six jobs. The professions hardest hit are forecast to be those that employ office workers. Forrester, however, notes that about half of the job destruction will be matched by job creation equal to 9 percent of today's jobs. The study points to "robot monitoring professionals," data scientists, automation specialists, and content curators as the biggest sources of new tech-related jobs. On net, Forrester forecasts that the impact will be a loss of 7 percent of jobs. That is still one out of every fourteen jobs. A recent World Economic Forum study, which is based on a survey of high-level corporate human resource types, put the number much lower. It predicts that in the next

8. Specifically, 60 percent of jobs are in occupations where at least 30 percent of the job is automatable using proven technology according to McKinsey Global Institute in "A Future That Works: Automation, Employment, and Productivity," January 2017.

9. Forrester, "Robots, AI Will Replace 7% of US Jobs by 2025," *Forrester.com*, June 22, 2016.

few years, only seven million workers worldwide will be replaced by automation.[10]

A survey from Japan has a very different set of findings. The survey, arranged by the research arm of the country's widely respected Ministry of Economy and Trade and Industry, posed a simple question: "What do you think about the impact of AI and robotics on the future of your job?" The possible replies were: 1) "I might lose my job," 2) "I don't think I will lose my job," and 3) "I don't know."[11] The may-lose-my-job responders accounted for about a third of the respondents overall. That's a lot in a tech-savvy society which has seen much more rapid automation and introduction of robots than we have seen in Europe and the US. The response, however, was much stronger among younger workers. Forty percent of those under thirty thought they might lose their job to a robot, while only 20 percent of those over sixty thought the same.

In 2014, Pew did interviews with over 1,800 tech experts, asking the million-dollar question: "Will networked, automated, artificial intelligence (AI) applications and robotic devices have displaced more jobs than they have created by 2025?"[12] The experts were in two camps, but before we get to that, here is the key message. Almost all the experts expected substantial job displacement by AI. What they differed on was whether there will be equally impressive job replacement.

About half the experts felt there will be significant net blue- and white-collar job displacement, which will lead to social upheaval, such as mass unemployability, vastly greater inequality, and breakdowns in social order. The other half were more optimistic. They had faith that humans' ingenuity will create masses of new jobs.

10. World Economic Forum, "The Future of Jobs Employment, Skills and Workforce Strategy for the Fourth Industrial Revolution," January 2016.

11. Masayuki Morikawa, "Who Are Afraid of Losing Their Jobs to Artificial Intelligence and Robots? Evidence from a Survey," RIETI Discussion Paper 17-E-069, 2017.

12. Pew Research Center, "AI, Robotics, and the Future of Jobs," August 2014.

If history is a guide, new occupations will appear and these will create many posts. There is, however, another way in which new jobs may be created and that is by digitech itself.

NEW JOBS DIRECTLY CREATED BY DIGITECH

There are at least three ways in which the breakneck advance of digital technology is creating jobs at an equally breakneck pace. The first has to do with the explosion of data. As more people get online and as we all do more online, the demand for online and phone-based services is exploding. Moreover, online activity is creating mountains of data. The size of the digital tsunami is amplified by the so-called internet of things, which means machines talking to machines online.

The only way to deal with this absolutely colossal wave of data is to employ white-collar robots. Since advanced AI, like Amelia and her "cobots," can't handle really unusual cases, humans will still be needed. Thus there will be a lot of substitution of AI for humans, but since the amount of work is exploding, the number of humans employed in such operations will expand. Here AI shouldn't be viewed as a straight-out job destroyer since, indeed, the only alternative to employing AI would be to ignore the data (as is often the case even today). "People who worry about job losses to automation tend to overlook the unprecedented data explosion businesses are experiencing, now accelerating out of knowledge workers' control and demanding automation to deal with it," write London School of Economics professors Leslie Willcocks and Mary Lacity.[13] Many of the firms that the professors studied have already adopted RPA solutions, and yet they have promised their workers that the robots would not lead to any layoffs—even if the RPAs meant that there would be no new hires in the department.

13. Mary C. Lacity and Leslie Willcocks, "What Knowledge Workers Stand to Gain from Automation," *Harvard Business Review*, June 19, 2015.

A British utility, studied by Willcocks and Lacity, "hired" more than three hundred RPAs to wade through three million transactions per quarter. They estimated that it would have taken six hundred people to do the same work manually. These synthetic workers didn't take any jobs at all—they simply allowed a company to make some money on the avalanche of information.

This sort of assurance cheered the workers and made it easier to train and integrate these "digital assistants." The workers embraced the newcomers because they viewed the bots as relieving them of the drudgery, thus leaving them more time to deal with idiosyncratic cases.

The second way digitech is directly creating jobs has to do with a curious feature of digital products—they are often free.

There are many striking differences between the mechanical automation that marked factory and farm jobs and the electronic automation that is hitting the service sector today. One is the price. Since it is almost costless on the margin to run white-collar robots—they are, after all, just computer programs—the price of the things they do is often zero. A whole slew of new services are free. Things we would have paid good money for—say Google Maps, TripAdvisor, and news sites—are often free in today's world. And free creates its own demand. Many services that would have involved lots of people, and therefore would have been expensive, are now offered for free, and we are "buying" these new services in a massive way. Examples include: digital pill reminders, CVS telemedicine, and robotic financial investment advice.

Rachel at Bank of America, Alexa at Amazon, and Apple's Siri make it almost free to ask for information, so we are asking for mountains of it. The result is that these firms are hiring. The basic reasoning is as easy as one, two, three: 1) AI software makes it feasible to charge a zero price to consumers for services that a few years ago would have been expensive; 2) people start using these services like crazy; and 3) the companies providing the new services hire people to look after the robots and do more human chores like management, accounting, human resource management, and the like.

A third way AI automation is creating jobs in rich nations is by reshoring back-office jobs that had been offshored to countries like India.

The idea of replacing high-cost workers doing routine manipulation of information that can be sent down a wire is an old one. Since the 1990s, many companies have sent these jobs overseas. This created a whole industry called business process outsourcing (BPO) that is today dominated by companies like Infosys.

RPA is good at many of the tasks that BPO companies now do. The cost savings are almost coercive. According to Genfour, which was acquired by Accenture in 2017, "While an onshore FTE [full-time equivalent worker] costing $50K (total cost) can be replaced by an offshore FTE for $20K, a digital worker can perform the same function for $5K or less—without the drawbacks of managing and training offshore labor."[14] Since the AI software cannot handle all cases, bringing back-office jobs back to America and Europe will create some jobs for white-collar humans along with lots of jobs for white-collar robots.

Another example of rapid job creation is the mass hiring that Amazon is doing. But here the distinction between net and gross job creation matters. To paraphrase the old saying, you can't make a blanket longer by cutting a foot of cloth off the top of the blanket and sewing only a half of a foot of cloth back on to the bottom. The rapid introduction of AI-trained robots into the workplace boosts productivity per worker, and this tends to reduce the number of workers needed. But by making things cheaper and quicker, robots are also increasing sales. Amazon provides a great example of this productivity-production foot race.

The Amazon Example—Trimming the Blanket

Amazon has deployed an army of white-collar robots to speed up what they call the "click to ship" time—the amount of time that elapses between the time you hit the "buy" button on your screen and the time the item actually leaves the Amazon warehouse closest to you.

14. Rita Brunk, "The ABC of RPA, Part 5: What Is the Cost of Automation and How Do I Justify It to the Leadership Team?" *Genfour.com*, July 21, 2016.

This automation has meant faster Amazon delivery, which in turn is helping Amazon and other online retailers undercut brick-and-mortar stores. With this e-commerce booming, Amazon is hiring. In 2017, almost a million people worked in warehousing in the US—that's up by over four hundred thousand workers, according to Bloomberg.[15] In the UK alone, Amazon created 2,500 new permanent jobs in 2016. In summer of 2017, Amazon announced it was looking for fifty thousand more workers.

For Amazon, AI automation radically reduced cost and improved timeliness. While this meant fewer workers were needed for a given pile of work, the better service meant a much larger pile of work and therefore more jobs at Amazon. Of course, the job creation by Amazon has implications for the number of jobs in traditional retail stores.

Much of the business that is going to Amazon is coming from traditional retail stores. And since Amazon is so much more efficient, the shift from in-store sale to online sales is reducing the number of jobs on net. Malls across the US are shuttering, and the impact on high-street stores in Britain is starting to be felt. In short, Amazon's new jobs are not net job creation.

The example of Amazon shows that the practical details matter. As the old saying goes: the difference between theory and practice is different in theory than it is in practice. That explains why it is insightful to turn to actual practice, namely service-sector occupations where robots are displacing workers today.

REALITY CHECK—JOBS BEING AUTOMATED TODAY

The world is a complicated place, so it helps to figure out what matters and what doesn't. It may well be that AI will cut in half the number of radio operators but since there are only 870 of them in the US, who cares?[16]

---

15. Patrick Clark and Kim Bhasin, "Amazon's Robot War Is Spreading," *Bloomberg*, April 5, 2017.

16. Bureau of Labor Statistics, "May 2017 National Occupational Employment and Wage Estimates."

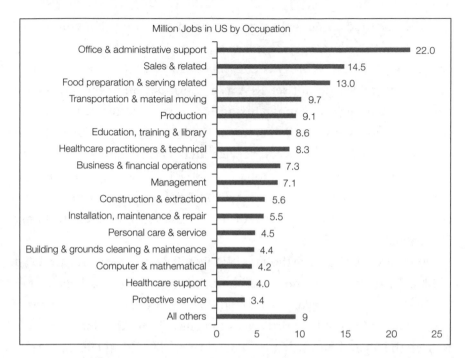

Figure 6.1 Millions of Jobs in US by Occupation, May 2016.
SOURCE: Author's elaboration of BLS online database.

Figure 6.1 shows which occupations in the US you should really care about since so many people work in them. The biggest category of all is the twenty-two million office workers. Many of them do things that AI can replace easily.

Office Work Automated

RPA is automating away many jobs in which workers are basically processing information and sending it on down an information assembly line.

It is hard to estimate how many of the twenty two million US office jobs RPA will eliminate, but the trend has spread across the developed world. The title of a 2016 KMPG report says it all: *From Human to Digital: The Future of Global Business Services*. KMPG's survey, which covered hundreds of global service companies, found that companies are turning to technology to replace human workers. Specifically, they are looking to RPA. KMPG is convinced that this will have an enormous impact. "We do not see RPA as a continuation of the large-scale automation of legacy manufacturing processes. Rather, it is a watershed, as there is no parallel that has the potential to reduce human workforce costs across every service delivery role."[17] Their survey found firms in European nations with strict employment protection laws to be especially interested in RPA. Of European firms, 80 percent were interested in this form of AI automation; the figure was only 50 percent in the US.

When asked how fast he thought RPA would displace workers, the head of Blue Prism, Jason Kingdon, was blunt: "My prediction would be that in the next few years everyone will be familiar with this. It will be in every single office." The stock market seems to believe him. Kingdon's company was worth £50 million when it went public in early 2016 and its share price has risen by 650 percent since then.[18]

The second biggest category of US jobs shown in Figure 6.1 is "sales and related occupations" with 14.5 million US workers.

Automation of "Walking Worker" Service Jobs

Automation in the service sector is not limited to software robots replacing brain workers. It is also coming to what we might call "walking service worker" jobs, that is, jobs that involve people walking around and manipulating physical things. The robots replacing these workers are not like Amelia and RPA; they are "steel-collar" robots—physical machines that move.

17. KPMG, *From Human to Digital: The Future of Global Business Services*, 2016.

18. Ian Lyall, "Small Cap Ideas: Could Blue Prism Be the Next Big British Software Champion with Its Robot Clerks?," *ThisIsMoney.co.uk*, March 21, 2016.

The Retail Sector

Retail stores are no strangers to automation. Self-checkout terminals have already replaced many workers in a whole range of shops. Some humans are still needed to handle unusual cases, but stores hire fewer people for checkout. A series of innovations are pushing the automation even further. Some US stores have apps that let customers get information on products via their mobile devices by scanning the bar code or taking a picture. This means fewer shop assistants.

Other stores are using AI to make the store shelves "smart." They use something called "proximity beacons" to send messages to shoppers' phones when they are near an item of special note. It can also enable a somewhat spooky, we-know-where-you-are sort of thing, like personal discounts on nearby items. Nordstrom uses one and Walmart is trialing one called iBeacon based on Apple technology.

US retail giant Kroger, which is the number two retailer after Walmart, introduced a new type of shelving whose edge (the narrow part that faces consumers) is digital. This is like a programmable video screen that uses sensors and analytics to provide buying recommendations, custom pricing, and detailed product information to customers. Again, this means better customer service with fewer employees.

Jobs are also being replaced on the inventory side of retailing. The US home improvement and appliance retailer Lowe's has introduced LoweBot. This is a free-ranging, self-driving robot that answers simple customer questions and can help them find products. Shoppers can type queries into its touch screen, or just ask. It speaks and understands English, Spanish, and a couple of other languages.

The five-foot-tall, rather bland looking robot also helps with inventory. The machine, which is basically a touchscreen on wheels with lots of sensors attached, can automatically scan the shelves and identify the goods in real time. LoweBot debuted in Silicon Valley stores in 2017. A competitor is the robot Tally, which patrols supermarket aisles when the store is open checking that all the products are in stock, correctly placed, and correctly priced.

The high-end department store Bloomingdale's started equipping fitting rooms with wall-mounted screens in 2017 that let customers scan things they are trying out to see if other colors or sizes are in stock. The system can also suggest other pieces in case the shopper wants to complete the look. These amenities make for a better shopping experience with the same number or fewer shop assistants.

These developments are so new that there is no research or data on job displacement, but the intent is absolutely clear. They are direct substitutes for humans. Machine learning has also been applied to physical jobs outside of factories.

Construction Jobs Automated — SAM the Bricklaying Robot

For people with a strong back but not much education, construction is one of the best jobs on offer. But this too is being automated. The New York firm, Construction Robotics, rents a robot called SAM (semi-automated mason) to US construction companies for $33,000 a month. SAM works with human masons (funny how the "human" in "human mason" would have been redundant in 2014). Here's how it works.

A conveyor belt delivers bricks to a robotic arm which then spreads mortar onto the brick and places it on the wall using laser sensors to get the placement just right. Humans are needed to load the bricks on the conveyor, shovel mortar into the hopper, smooth off excess mortar, and control the whole system with a tablet computer. SAM lays 1,200 bricks a day, two to four times more than a human bricklayer.

Construction Robotics reckons that SAM cuts labor costs for bricklaying projects by roughly 50 percent. This means fewer bricklaying jobs per construction site, but SAM will not eliminate the bricklaying profession. Those who keep their jobs will be more productive; those who lose their job to SAM will have to find something else to do.

Like construction workers, security guards tend to have high school educations and a sturdy disposition. Their jobs are also under threat.

Security Guards

While having a security guard around is very useful just in case something bad happens, the main task of security guards is just being there—and being capable reacting if something bad does happen. But exactly because there is a security guard on hand, bad things are less likely to happen. This paradox—that guards are not typically needed when they are there—has encouraged automation.

One Californian company, Knightscope, leverages the mismatch by providing robot security guards who can do the "being there" part while staying continuously in touch with real human security guards who can take over if a real incident occurs. Knightscope guards are already used in malls and out on the streets of San Francisco, where it chases away homeless people. It has cameras, laser scanners, a microphone, and a speaker. It can drive itself around at a slow walking pace.

It is not a good as a human security guard, but it is a whole lot cheaper, renting out at seven dollars an hour (below minimum wage). And it doesn't need breaks or overtime on holidays. Still it has its flaws. One robot patrolling a mall in Washington, DC, rolled into a fountain and drowned itself in 2017.

Lower down the service-sector food chain, so to speak, are food preparation jobs, which often pay minimum wage. Almost one in eleven US workers are involved in food preparation and food serving: thirteen million jobs.

Food Preparation Jobs Being Automated

McDonald's and other big US chains like Chili's Grill & Bar, Applebee's, and Panera Bread are automating some tasks—taking some of the work out of workers, so to speak. One practice that is spreading rapidly is the use of touchpad tablets to take orders directly from customers.

Typically installed at each table, tablets reduce the number of workers each restaurant needs. It also means that people don't have to wait for their waiter (ever wonder where the "wait" in waiter comes from?). And strangely enough, these devices induce people to order more.

Whether it's the guilt avoidance from not having to pronounce out loud, "yes, I'd like the chocolate ice cream for dessert," or just the convenience of spontaneous ordering, the amount per check is higher for waiterless orders, according to research by one of the ordering-tablet makers, Ziosk. The trend is growing: Ziosk has already shipped hundreds of thousands of such tablets.

Restaurant automation is also coming via smartphones. The historic maker of cash registers, NCR (it stands for National Cash Register), has leapfrogged itself by offering a app, NCR Mobile Pay, that allows restaurant customers to order, browse their bill, reorder menu items, call the waiter, tip and pay, and get a receipt by email—all via their smartphones.

Automation of restaurant kitchens is just starting. Take Flippy, a burger-making robot that is being developed in cooperation with the CaliBurger chain. Flippy, which is basically a robotic arm with sensors wielded onto a cart, can roll up to any standard grill or fryer and start cooking just like any minimum-wage worker. No redesign of the kitchen is necessary.

Flippy unwraps the pre-made burger patties that all fast-food kitchens use, slaps them on the grill, and flips them when the time comes—all using thermal sensors, cameras, and its onboard AI program. It can integrate into the restaurant's system and take orders directly from the customer counter. So far, Flippy still needs humans in the loop (to apply the cheese and other toppings), but a company called Momentum Machines created a machine that would eliminate all the food preparation jobs.

"Our device isn't meant to make employees more efficient, it's meant to completely obviate them," asserted Momentum Machines cofounder Alexandros Vardakostas in 2012.[19] The company's robot, which is about the size of a small walk-in refrigerator, takes in raw food and spits out wrapped and bagged burgers at a maximum rate of about a hundred per hour. That was in the early days.

19. Lora Kolodny, "Meet Flippy, a Burger-Grilling Robot from Miso Robotics and CaliBurger," *SingularityHub.com*, March 7, 2017.

Perhaps realizing that this sort of brash, anti-job sentiment might not go over well, Vardakostas changed his tune when he opened his first automated burger joint in June 2018: "Our utopian future is one where there is more creativity and more social interaction, while staff members also get to be more creative and social."[20] The company, re-branded as Creator and now supported by Google Ventures, is clearly trying to get ahead of any backlash that radical automation might cause among customers and workers. The plan is that they will pay employees well above the minimum wage and allow them to spend 5 percent of their time reading educational books of their choice.

The economics of fast-food automation are being accelerated by the rise of minimum wages in some US states. As the former CEO of McDonald's USA, Ed Rensi, put it bluntly: "It's cheaper to buy a $35,000 robotic arm than it is to hire an employee who's inefficient and making $15 an hour."[21]

Robots have also started to elbow their way into the pizza business. A San Francisco Bay Area start-up, Zume Pizza, uses a robot—or as they call it, a "doughbot"—to shape dough into perfect pizza crusts in seconds. Other robots spread the sauce and pop the pie into the oven. You order the pizzas online with your smartphone. There is no counter and no store front.

Zume produces more than two hundred pizza pies per day with only four people in the kitchen. They plan to reduce the number of workers with more robots and more AI. If their plans work out, "it would be like Domino's without the labor component," says co-CEO Alex Garden. "You can start to see how incredibly profitable that can be."[22] Zume spends just 14 percent of revenue on workers, compared to 30 percent for Domino's.

20. Quote in Melia Robinson, "This Robot-Powered Burger Restaurant Says It's Paying Employees $16 an Hour to Read Educational Books while the Bot Does the Work," *Business Insider*, *UK.businessinsider.com*, June 22, 2018.

21. Quote in Julia Limitone, "Former McDonald's USA CEO: $35K Robots Cheaper Than Hiring at $15 Per Hour," *FoxBusiness.com*, May 24, 2016.

22. Sarah Kessler, "An Automated Pizza Company Models How Robot Workers Can Create Jobs for Humans," *QZ.com*, January 10, 2017.

Transportation Jobs

Something like one in fourteen US workers is involved in transportation of some type. That's about ten million jobs, with about half of them driving some sort of vehicle. These jobs are on their way to automation as many know. Indeed, these are probably the service-sector jobs where the threat of service-sector automation is most widely discussed.

Self-driving trucks and cars are a reality, but it is not yet clear how fast the technology will take off. As David Rotman of *MIT Technology Review* magazine observes, "any so-called autonomous vehicle will require a driver, albeit one who is often passive. But the potential loss of millions of jobs is Exhibit A" in the threat AI poses to service-sector jobs that were previously considered safe from automation.[23]

A report by President Obama's White House economists and science advisors, *Artificial Intelligence, Automation, and the Economy*, estimates that automated vehicles could threaten 2 to 3 million US jobs. Many of these workers, including the roughly 1.7 million truck drivers, are some of the best jobs available to people without advanced education.

Actually implementing the automation will not be easy or smooth given how regulated these industries tend to be—at least in part due to the safety issues posed for the general public. It is easier to imagine a future when all vehicles are automated and they coordinate with each other. The hard part is when some are driven by humans and others by robots.

But automation is not limited to unskilled jobs in the service sector. Doctors, lawyers, journalists, accountants, and many other professionals make good money because they have mastered masses of information and garnered years of experience in applying it to new situations. That, however, is exactly what AI does extremely well. If you replace "experience" with "data"—so experience-based pattern recognition becomes data-based pattern recognition—you have a pretty good description of

---

23. David Rotman, "The Relentless Pace of Automation," *MIT Technology Review*, February 13, 2017.

the activities where machine learning has or soon will be better than the average human. This is already happening in medicine.

Medical Jobs

Healthcare is a very large sector. In the US, about 12 million people work in the industry. Only one out of twenty of these are doctors; nurses make up one in five. The UK's National Health Service directly employs 1.5 million. Much of healthcare is fairly routine, but almost all of it turns around experience-based pattern recognition. This puts it squarely in the path of advancing AI.

White-collar robots are good and getting better at processing images and patient history information. They are already used in making diagnoses. Yet instead of replacing doctors, white-collar robots are acting as yet another diagnostic device that doctors employ in doing their jobs. Some of the more innovative uses of white-collar robots are in psychology.

Ellie is an on-screen white-collar robot (some call it an avatar but that is focusing on the image and underplaying the technology driving the image). She looks and acts human enough to make people comfortable talking to her. Computer vision and a Kinect sensor allow her to record body language and subtle facial clues that she then codifies for a human psychologist to evaluate. Research shows that she is better at such data gathering than humans—in part because people feel freer to open up to a robot.

University of Southern California researchers created Ellie as part of a program financed by the US Defense Advanced Research Projects Agency. The program's aim is to help veterans with post-traumatic stress disorder (PTSD). "One advantage of using Ellie to gather behavior evidences is that people seem to open up quite easily to Ellie, given that she is a computer and is not designed to judge the person," explains her co-creator, Louis-Philippe Morency.[24] Other robo-pychology

24. Nathan Jolly, "Meet Ellie: The Robot Therapist Treating Soldiers with PTSD," *News.com.au*, October 20, 2016.

applications help provide therapy to patients. Woebot, for example, engages people in daily conversations to help them with mental health issues. Mostly, it asks questions that encourage the user to reformulate negative thoughts in more objective ways. Robo-medicine is also in common use in hospitals.

In Singapore's Mount Elizabeth Novena hospital, IBM's Watson is used to monitor patients' vital signs in place of human nurses. The hospital's CEO, Louis Tan, notes that Watson is just an aide: "It doesn't mean nurses are absolved of responsibility. It just means they have another aid. It's more efficient and safer for the patients."[25] Another labor-saving form of automation is aimed at reducing the time doctors spend on routine things.

"A lot of visits to the general practitioner (as many as three in five) are for minor ailments, advice or things that you could sort out yourself with over the counter medicines," notes Matteo Berlucchi, who is chief executive of Your.MD, which produces a medical white-collar robot. This is a smartphone app that mimics a consultation with a general practitioner. "It's not a matter of replacing doctors," says Berlucchi, but rather "taking some of the easier and more mundane situations off the hands of real doctors and having AI sort them out."

This is basically "pre-primary care" that helps people who aren't feeling well decide whether they need to see a doctor. The UK's National Health Service sees the potential and has approved the information that the app uses. There are more spectacular examples of robo-medicine.

In 2016, Japanese doctors consulted Watson after their treatment failed. As it turned out, the patient—whom doctors had diagnoed with acute myeloid leukemia—was suffering from something else. Watson consulted its database of twenty million cancer research papers, looking for patterns that matched the patient's genes and medical records. Based on the patterns it recognized, it guessed that she was suffering from a rare form

---

25. Quotes from Jeevan Vasagar, "In Singapore, Service Comes with a Robotic Smile," *Financial Times*, September 19, 2016.

of leukemia that the human doctors hadn't considered. This took the robot ten minutes.

Once Watson proposed the new diagnosis, the doctors decided the robot was right and changed their treatment. This probably saved the woman's life. Note that Watson did not replace any doctors in this case. It is easy to imagine, however, that Watson could allow one doctor to provide a given pile of medical services in less time. Watson is thus a form of automation. But also note that if it became widely used, it would involve a reverse "skill twist." Watson would be a replacement for the most specialized, highest-paid cancer doctors, but it would be a better tool for average doctors. This is a classic example of AI upskilling average workers.

Pharmacies Automated

Counting pills takes up a lot of pharmacists' time. The University of California San Francisco Medical Center, for example, has about six hundred patients at any one time that take an average of ten different medications each. That occupies a couple hundred pharmacists and pharmacy technicians, but it would require far more were it not for a pill-picking robot called PillPick. This robot picks, packages, and dispenses individual pills. In many cases, it adds a barcode to provide extra assurance that the right patient gets the right medication.

As is often the case when humans offload routine tasks to robots, consistency has risen with PillPick. Andrew Zaleski, writing on CNBC.com in November 2016, notes that a study at a Houston hospital found five errors for every 100,000 prescriptions filled by human pharmacists.[26] It was just such an error that pushed the Medical Center toward automation. "A nurse made an error of putting the decimal point in the wrong place and we overdosed a patient, and at that point, we made a commitment that we didn't ever want that to happen again," said the Center's chief executive

---

26. Andrew Zaleski, "Behind Pharmacy Counter, Pill-Packing Robots Are on the Rise," *CNBC. com*, November 15, 2016.

officer, Mark Laret. The robot filled about 350,000 prescriptions during its probationary phase-in period—all without errors.

Journalism Automated

The *Washington Post* has a fantastically productive journalist who produced over five hundred articles in the days following the November 2016 US elections; every House, Senate, and gubernatorial election was covered in real time. The reporter's name is Heliograf, and he is a robo-reporter. The newspaper's sixty human political reporters focused their attention on the high-profile, dramatic, or close contests. Heliograf, like a robo-intern, was left the dreary job of reporting on the outcomes of the less sexy contests.[27]

In the 2012 election, by contrast, the *Washington Post* assigned four human reporters to getting out stories on the out-of-the-way results. In twenty-five hours, they managed to cover only a small fraction of the races Heliograf wrote about.

This automated election reporting has also been used in France. Working with *Le Monde* during France's 2015 election, an IT company used automated writing software to produce text for 150,000 web pages in four hours. The IT company's CEO, Claude de Loupy, notes: "Robots can't do what journalists do, but they . . . can do amazing things, and it's a revolution for the media."[28] Many other news organizations, like AP News Service, are using commercially available robo-writing software.

But how good is robo-writing? The US's equivalent of BBC, National Public Radio (NPR), staged a man-versus-machine duel—somewhat like the chess match in 1997 pitting world chess champ Gary Kasparov against IBM's Deep Blue computer. This time, it was NPR White House

---

27. The information on the *Washington Post* is drawn mainly from Joe Keohane, "What News-Writing Bots Mean for the Future of Journalism," *Wired.com*, February 16, 2017.

28. Damian Radcliffe, "The Upsides (and Downsides) of Automated Robot Journalism," *MediaShift.org*, July 7, 2016.

correspondent Scott Horsley versus a robo-writer called WordSmith. The news event trigger was to be the earnings report for the fast-food company Denny's. The output was to be a short radio story. The machine took two minutes to finish; the human took seven. The judges, NPR listeners voting online, thought the human's story was richer and more engaging.

Is robo-journalism displacing human journalists? The mood in the *Washington Post* newsroom is, so far, pretty positive. Although they have not given the robot a cute name, there is acceptance. The union representative, Fredrick Kunkle, said: "We're naturally wary about any technology that could replace human beings, but this technology seems to have taken over only some of the grunt work."[29]

As already mentioned, some legal jobs are also under threat.

Legal Work Automated

In late 2016, JP Morgan's AI software, COIN, automated the reading and interpretation of commercial loan agreements. Before COIN, the work cost an estimated 360,000 hours by lawyers and loan officers. Now it's done much faster and with fewer errors by a system that never sleeps while reading through 12,000 or so contracts a year. Plans are afoot to use COIN for complex legal filings like credit-default swaps and custody agreements.

In a refrain that is almost regulatory by now, JP Morgan's chief information officer, Dana Deasy, asserts that COIN doesn't eliminate jobs. It just frees up the lawyers and loan officers for better things. "People always talk about this stuff as displacement. I talk about it as freeing people to work on higher-value things, which is why it's such a terrific opportunity for the firm."[30] That may be true for the high-end lawyers, but there are about a million people working in legal services in the US. Many of the

29. Quotes from Joe Keohane, "What News-Writing Bots Mean for the Future of Journalism," *Wired.com*, February 17, 2017.

30. Quoted in Casey Sullivan, "Machine Learning Saves JPMorgan Chase 360,000 Hours of Legal Work," *Technologist* (blog), FindLaw.com, March 8, 2017.

things they do today are, or soon will be automatable. "The legwork of the legal industry is reading documents," notes Jan Van Hoecke, co-founder of the legal AI start-up RAVN, and his company "is about automating the reading process." The company's AI reads and interprets documents—extracting information faster and more accurately than humans. It is already widely used among top law firms and increasingly by corporate legal departments.[31]

One area where technology substituting for young lawyers burning the midnight oil is what lawyers call "discovery." That's the part—which you've seen in countless courtroom dramas—where the smart young things plow through stacks of documents to find evidence that will exonerate their client or incriminate the other side's evildoer. Much of this is now done by AI-charged, white-collar robots.

On the lighter side is a legal-bot, called DoNotPay. It's a computer program, accessible for free online, that uses Facebook Messenger to interview you about your traffic tickets. It then instantly spits out legal advice and documents showing how you could beat the ticket.

It was created by a very interesting young British man. "When I started driving at 18, I began to receive a large number of parking tickets and created the DoNotPay as a side project. I could never have imagined that just over a year later, it would successfully appeal over 250,000 tickets." According to an interview in Forbes, Joshua Browder, who taught himself computer programming at the age of twelve, only worked on DoNotPay between midnight and three in the morning.[32]

He is now a twenty-something studying law at Stanford University. An idealist at heart, Browder adapted the robo-lawyer to help US and

---

31. Deloitte's 2016 report titled *Developing Legal Talent: Stepping into the Future Law Firm* suggests that something like two-fifths of legal jobs in the US may be automated in the next two decades. Another study suggests that existing AI could replace one in eight hours of legal work done in the US (Dana Remus and Frank Levy, "Can Robots Be Lawyers? Computers, Lawyers, and the Practice of Law," *SSRN.com*, December 11, 2015.)

32. Alexander Sehmer, "A Teenager Has Saved Motorists over £2 Million by Creating a Website to Appeal Parking Fines," *Business Insider UK*, December 30, 2015.

Canadian refugees complete immigration forms. In the UK, it helps asylum seekers get financial support from Her Majesty's Government.[33]

Another high-end profession where jobs are being axed is financial services.

Finance

Many people these days manage their own money to some extent, and almost everyone is having to take more responsibility for big financial decisions like retirement. Basic information about financial realities, however, is still difficult to come by. Talking to a banker or financial advisor can be expensive, and many are really just salespeople trying to earn commissions.

A new trend in personal finance is to use white-collar robots for these things. UBS, for example, has hooked up with Amazon's Alexa to deliver answers to simple financial queries. The US government-sponsored mortgage company, Fannie Mae, has replaced teams of report-writing financial analysts with white-collar robots. This allowed the company to review performance quarterly instead of annually and to cover far more borrowers.

The leading investment bank, Goldman-Sachs, has automated many trading desk jobs. In 2000, the company employed six hundred traders in its New York office. Now there are just two traders working with two hundred computer engineers. In its foreign exchange trading unit—which used to be dominated by high-paid, high-finance types—a third of the staff are now computer geeks (and the total head count is way down). The impact can be good for those at the top. Babson College professor Tom Davenport says, "The pay of the average managing director at Goldman will probably get even bigger, as there are fewer lower-level people to share the profits with."[34]

33. Quotes from Megha Mohan, "The 'Robot Lawyer' Giving Free Legal Advice to Refugees," *BBC Trending* (blog), March 9, 2017.

34. Quoted in Nanette Byrnes "As Goldman Embraces Automation, Even the Masters of the Universe Are Threatened," *TechnologyReview.com*, February 7, 2017.

The examples are endless and growing since many jobs in finance involve doing things that white-collar robots are really good at, namely—making fast decisions based on tons of data. And this job displacement could go much further.

Marty Chavez, Goldman's deputy chief financial officer notes that investment banking is in for the globot treatment. Investment bankers involved in mergers and acquisitions earn, on average, $700,000 a year, so the profit motive for slimming the numbers is clear. While many of the skills—like selling ideas and building relationships—will stay with humans, the company has identified over a hundred specific tasks that could be automated.

In 2018, former Deutsche Bank chief executive John Cryan guessed that that up to half of the German bank's workforce could be replaced by technology. As Barclays investment bank CEO Tim Throsby said, "If your job involves a lot of keyboard hitting then you're less likely to have a happy future." Amplifying the point, Richard Gnodde, head of Goldman Sachs International, said: "There are so many functions today that technology has already replaced and I don't see why that journey should end any time soon."[35]

WHERE IS ALL THIS HEADING?

Globots—and that means globalization in the shape of telemigrants and cognating computers in the form of white-collar robots—are driving a new transformation. This new version of the old disruptive duo—automation and globalization—will not be gentle. Many occupations that were sheltered from the duo are now being subjected to both automation and globalization. Many of these jobs are in offices and the results will be rather grim.

35. Quotes from Laura Noonan, "Citi Issues Stark Warning on Automation of Bank Jobs," *Financial Times*, June 12, 2018.

These changes won't eliminate many occupations—since most work activities include some things that neither white-collar robots nor telemigrants can handle. But the Globotics Transformation will surely lower the headcount in many of today's most common service-sector occupations. Digitech is also creating some jobs, but indirectly and generally only for workers with very specific skills.

This means that the disruption, displacement, and dismay that has been experienced by factory workers since 1973 will soon be shared by many white-collar workers. Given the rapacious rate of digitech progress, these changes will disorder professional and service-sector jobs radically faster than globalization disrupted the manufacturing sector in the twentieth century and agricultural sector in the nineteenth century.

If history repeats itself, the rapid innovation will lead people into jobs that remain sheltered, but in the meantime, things could get mean. There will be an upheaval. There will be a backlash.

7

The Globotics Upheaval

Bill Gates is worried that digitech will launch an upheaval. This should worry all of us. Gates can't know the future—that's unknowable—but he has proved time and again that he understands what digital technology can do. He became one of the world's richest men by guiding Microsoft through decades of "holy cow" moments.

In Gate's view, job displacement is coming too fast for the economy to absorb. "You cross the threshold of job replacement of certain activities all sort of at once. You ought to be willing to raise the tax level and even slow down the speed."[1] And Gates is not the only rich tech guy who's worried.

The technology entrepreneur, Elon Musk, who owns rocket ships as a sideline to being CEO of Tesla, also knows a thing or two about disruptive technologies. Tesla was valued more highly by the stock market in 2017 than any of the traditional carmakers. And Musk is as concerned as Gates. Here is how he phrases it: "What to do about mass unemployment? This is going to be a massive social challenge. There will be fewer and fewer jobs that a robot cannot do better. These are not things that I wish will happen. These are simply things that I think probably will happen."[2]

1. Quote from Kevin Delaney, "The Robot That Takes Your Job Should Pay Taxes, Says Bill Gates," *Quartz*, February 17, 2017.

2. Quote from Quincy Larson, "A Warning from Bill Gates, Elon Musk, and Stephen Hawking," *freeCodeCamp.org*, February 18, 2017.

The CEO of Amazon, Jeff Bezos—another successful surfer of technology waves—says: "It's probably hard to overstate how big of an impact it's going to have on society over the next twenty years."[3] Devin Wenig, who is the CEO of eBay points out: "While the promise of AI has been known for years, the current pace of breakthrough is stunning. Machines are set to reach and exceed human performance on more and more tasks, thanks to advances in dedicated hardware, faster and deeper access to big data, and new sophisticated algorithms that provide the ability to learn and improve based on feedback."

The late Stephen Hawking never knew much about business, but as one of the world's most eminent physicists, he was well placed to judge the future course of digitech. He warned: "The automation of factories has already decimated jobs in traditional manufacturing, and the rise of artificial intelligence is likely to extend this job destruction deep into the middle classes, with only the most caring, creative or supervisory roles remaining."[4]

These rich guys have put their finger on the thing that will turn the Globotics Transformation into the globotics upheaval. Having a good job and belonging to a stable community are critical elements of a successful life in today's economy. Up till now, many of these "successful lives" were lived by people working in white-collar and professional jobs. And up till now such jobs were sheltered from both globalization and robots. Globots are changing that reality.

All change is associated with both pains and gains. But when change comes very quickly, people end up having to undertake "emergency maneuvers" that can be extremely costly personally, financially, and socially. That's why our governments almost always phase-in changes slowly. It gives people time to reorder their affairs in an orderly manner. The globotics upheaval, however, is not coming in an orderly manner.

3. Quoted in Walt Mossberg, "Five Things I Learned from Jeff Bezos at Code," *Recode* (blog), June 8, 2016.

4. Stephen Hawking, "This Is the Most Dangerous Time for Our Planet," *The Guardian*, December 1, 2016.

When tens or hundreds of millions of Americans, Europeans and other advanced-economy citizens are forced to change jobs, the transformation will—in any version of the future—produce economic, social, and political upheaval. But it's more complicated than that.

MISMATCHED SPEED AND THE UPHEAVAL

Transformative technology is as old as the sun, or at least as old as the sundial. In this sense there is nothing new about the Globotics Transformation, and there is nothing wrong with its direction of travel. Technological progress is a good thing and in any case it is irresistible.

The technologies that allow computers to think and allow foreign freelancers to work in our offices reside in software and on internet platforms. These are things that Western-style democracies have a very hard time controlling. That means that globots are coming to change our lives—at least eventually. Governments may slow the pace but they cannot stop it. In the long run, all will be for the best. The age of globots will make the world a better place—once the kinks are worked out. Globots will make us more productive and eliminate dull, repetitive work. They will, in a sense, allow human jobs to be more human-like. They will cut out all the robot-like things that people have to do today.

Upheavals, however, are never driven by what will happen in the future. They are driven by what is happening today. That's where the danger lies. The problem lies with the inhuman velocity of the changes, or more precisely, with the mismatch between the speed of job destruction and the speed of job construction. Digital technology is driving mass job displacement at a furious pace, but it is doing little to foster mass job creation. The point is straightforward.

Many of today's high-tech entrepreneurs are making billions (or hoping to) by replacing high-wage workers with lower-cost foreign freelancers, or even lower-cost white-collar robots. That's the business model—saving money by replacing workers in high-income countries. While the business people driving the job destruction are naturally reluctant to talk about it directly, AI scientists are not.

Job Destruction Is the Business Model

We should listen to Andrew Ng. He is one of the intellectual high priests of digital technology. He was the chief scientist at the Chinese online search giant Baidu, leading over a thousand researchers. Before that, he worked at Google developing the company's breakthrough machine-learning approach, called Deep Learning. This is the thing behind many of Google's wonders including its self-driving cars. As if all that wasn't enough for one person's career, when he was a professor at Stanford University, he co-founded the online education platform Coursera. His YouTube lecture on AI has been watched over 1.5 million times.

Ng is clear about the job-destroying aspects of digital technology. "I have so many friends working on significant projects that are squarely targeting many thousands or tens of thousands of people's jobs," Ng said. "These jobs are squarely in the bull's-eye." Speaking at the 2017 Consumer Electronics Show in Las Vegas, Ng ruefully adds in his American-Chinese accent with a slight Hong Kong heft: "And frankly those tens of thousands of people doing those jobs now have no idea that there are very serious projects underway that could automate a lot of those jobs."[5] Projecting forward, he says that if a human can perform a mental task in less than a second, it's likely that an AI computer can do the task faster, more consistently, and at a lower cost.

One of the leading providers of white-collar robots has a marketing pitch that brings home the intention point. Blue Prism refers to its suite of computer programs as "digital labor." On its website it announces: "multi-skilled software robots are implemented as digital labor in the most demanding enterprise back-office environments to eliminate the

---

5. Quotes from Adam Lashinksy, "Yes, AI Will Kill Jobs. Humans Will Dream Up Better Ones," *Fortune*, January 5, 2017.

disproportionately low-return, high-risk, manual data entry and processing work that humans shouldn't be doing."[6] These solutions have already been applied to the automation of back-office tasks in banking, telecoms, energy, government, financial services, retail, and healthcare.

The main point to keep in mind here is that the geniuses at Google, Amazon, Microsoft, Infosys, IBM, and so on are not working to create new jobs. They are working to displace them.

When it comes to the other type of globot—telemigrants—the mistmatched speed point is less clear as yet. Freelancing is booming but so far it mostly involves domestic workers, not telemigrants. The intentionality is also less clear. Profit motives are surely behind employers' ramping up their use of freelancers, but to date much of this has been creating jobs for domestic workers.

For example, the online payment company, Paychex, studied over 400,000 freelancers' resumes that were posted on Indeed.com (a job matching website). What they found was that "for the majority of the 1970s, '80s, and even '90s, working generally meant heading off to a typical 9-to-5 job. But during the new millennium, the freelance economy took flight. Between 2000 and 2014, freelance jobs listed on the resumes we examined increased by over 500 percent." The same is happening in Europe. From 2004 to 2013, the number of freelancers grew by 45 percent on average.[7]

An interesting driving force behind the trend is a concern—by workers—about the impact of white-collar automation on traditional 9-to-5 jobs. A large survey done by LinkedIn and Intuit in 2017 found this to be an important motive.[8] But this may be, as the old saying goes, "jumping out of the pan and into the fire." The trouble is that once companies arrange things to make it easy to hire domestic freelancers, there is little to

6. Alastair Bathgate, "Blue Prism's Software Robots on the Rise," *Blueprism* (blog), July 14, 2016.

7. Patricia Leighton and Duncan Brown, "Future Working: The Rise of Europe's Independent Professionals," EFIP Report, *Freelancers.org*, 2013.

8. Linkedin, "How the Freelancing Generation Is Redefining Professional Norms," *LinkedIn* (blog), February 21, 2017.

stop them from switching to lower cost foreign freelancers. As mentioned, the massive progress in machine translation, the rise of international freelancing platforms, and improved telecommunications is making telemigration a reality. As this catches on, the swapping foreign freelancers for domestic ones is likely to start snowballing.

Job creation is driven by a very different process.

Job Creation and Human Ingenuity

Some jobs are being created by digital technology as we saw before. Today's tidal wave of data is creating some new jobs for humans who are paid to make use of the data. The fact that new digital services are free is also a new source of new jobs even though much of the work behind free services like WhatsApp is done by white-collar robots. And digitech advances have also made it profitable to shift some service sector jobs that were previously done in India, for example, back to high-income nations.

But the number of such jobs is quite limited. Even at Alphabet—the wildly innovative and fast-growing company that owns Google—the net job creation between 2007 and 2017 was only 71,300 people.[9] That's just a drop in the US job-market bucket with its 140 million workers. And in any case, becoming a Googler is just not an option for most of the US hospitality workers whose jobs will be displaced by automation in the next few years.

The simple fact is that using digitech to create jobs is not the main focus of today's research and investment. Few companies are searching for ways to use digitech to create whole new categories of jobs. But there is no *technological* reason why digitech could not be used to do this.

White-collar robots with great diagnostic capabilities could, for instance, create a whole new class of medical professionals. People in this hypothetical occupation could do more than nurses, but less than doctors.

---

9. Statistics from *Statisa.com,* www.statista.com/statistics/273744/number-of-full-time-google-employees/f

Armed with Amelia-like digital assistants, men and women with far fewer years of training than a doctor could provide simple medical services. They could also be the medical profession's eyes-and-ears on the ground, identifying more severe cases that need the attention of doctors. They could help spread knowledge that prevents disease. We would all get better medical services at a lower cost.

There is no reason that this sort of intermediate occupation couldn't also work in other professions. AI could "upskill" workers with less education than lawyers, engineers, accountants, tax specialists, and investment advisors thereby creating masses of new "semi-professional" jobs. The new occupations would make all sorts of professional services more affordable and thus create new demand for the new services.

The catch is that creating new categories of occupations would require a sustained effort on regulatory and societal fronts. It would require new laws, new attitudes among customers, and acceptance from existing professionals. The job creation, in other words, would take a long time. It would not make anyone rich in the next five years.

The sad reality is that it is a lot easier and faster to make money by eliminating jobs than it is to make money by creating jobs. In short, there's nothing technologically inevitable about the mismatch in the speed of job destruction and construction, it's just about profits. And it is not forever.

Past economic transformations did not lead to permanent unemployment. When automation and globalization eliminated agricultural employment during the Great Transformation, new jobs were created in the industrial and services sectors. Likewise, the elimination of factory jobs from 1973 during the Services Transformation was accompanied by the creation of new jobs in the service sector.

Many of these new jobs were really new. During the Great Transformation, entrepreneurs invented many unheard-of products that turned out to sell well and they hired lots of workers to make them. During the Services Transformation, entrepreneurs invented new services that people were eager to pay for. Since most services involve people doing things for people, the new services created masses of new jobs. And as incomes rose, our demand for existing services swelled. We all started buying more medical, educational, and entertainment services for instance.

But what drove this invention and the resulting job creation?

The answer surely lies at least in part on new technical possibilities, but the hard part of creating something new is not the appearance of a new possibility. The hard part is finding the human ingenuity necessary to think up the new jobs. An even harder part is finding someone with the drive and entrepreneurship that can turn the ideas into realities.

Job creation, in other words, is limited by very human factors: things move at a pace that seems normal to our walking-distance brains, not at the explosive pace of digital technology. This matters because it means that we cannot count on new jobs appearing at anything close to the same rate that they are disappearing. There is no "Moore's Law" behind human ingenuity and entrepreneurship. Human ingenuity and entrepreneurship will do their job and find jobs for all of us eventually, but if history is a guide, that could take a long time.

When jobs are displaced at a breakneck pace but created at a leisurely pace, many people who thought they had stable, well-paying careers will find themselves struggling. This outcome has critical implications for the globotics upheaval. Remember how it played out for manufacturing workers during the Services Transformation from 1973. Many ex-factory workers found new jobs but often they were jobs that took them a whole step down the socioeconomic scale. The workers that globots lay-off in coming years will face many of the same bad choices that manufacturing workers did in recent years.

When it comes to white-collar robots and the automation of service jobs, the basic mismatched-speed point is well captured by a slight twist on the old (pre-DNA testing) Latin saying, "The mother is always certain, the father is never certain." When it comes the globotics upheaval, it should be "job displacement is always certain, job creation is never certain."

But how fast will it happen?

How Fast Will Job Displacement Outstrip Job Replacement?

How fast is not a question that can be answered with any precision. Think of it as hurricane forecasting. We know with certainty that there will be hurricanes in the Atlantic every year, and we even have a good idea of the months during which they will appear. But until a hurricane actually forms, it is impossible to know when and where it will cause disruption.

The deep reason is that weather is subject to all sorts of tipping points and accelerating feedback loops. Job displacement is governed by similar things, but with the added complexity of competition among existing companies, and between existing companies and yet-to-appear start-ups. This throws an unpredictable human element into the equation. Job creation is even less predictable since it will, as in the past, come in activities we can't even imagine today—and the unimaginable is something that is very hard to think clearly about. This brings us to Fiedler's main rule of forecasting: "give them a number or give them a date; never both."[10] Fiedler was also the one who said, "he who lives by the crystal ball soon learns to eat ground glass."

Fiedler's quips explain why technology and business experts are significantly more reluctant to pin down the timing of the job displacement than the number of jobs that are likely to be displaced. They are happy to give a number, but not a date. This is natural. It is just very hard to predict things since business transformation—and that's what globotics is doing—is not a hard science.

The Economist Intelligence Unit, for example, explains why so many companies were already investing so heavily in AI capabilities in 2016. "In time-honored business fashion, it is a combination of fear and hope. Competitive pressures are spurring companies on, and there is a sense

---

10. Edgar Fiedler served as Assistant Secretary of the Treasury for Economic Policy in the 1970s; these quotes are from Paul Dickson, *The Official Rules: 5,427 Laws, Principles, and Axioms to Help You Cope with Crises, Deadlines, Bad Luck, Rude Behavior, Red Tape, and Attacks by Inanimate Objects* (Mineola, NY: Dover, 2015).

of urgency amongst many industry managers about not falling behind."[11] Over a third of the CEOs they surveyed thought that digitech would allow new entrants to disrupt their business, so delaying would leave them vulnerable. When fear and competition come into play, especially when much of the change is likely to come from companies that don't even exist, precise predictions are problematic.

One very direct—but very partial—measure of the rapidity of job displacement is the swiftness with which the providers of robotic process automation (RPA) software solutions are growing. Blue Prism is the leading RPA provider.

Remember that the company sells software whose purpose is to reduce their human headcount in the service sector. The company's revenue at the end of 2017 was $25 million. Investment banks predict it will be $100 million by 2020, and $500 million just years after that.[12] Phil Fersht, of the specialized consulting group HfS, expects RPA software sales to grow at a compound annual growth rate of 36 percent—which means it will double every two years[13] The growth driven by a desire for cost savings and a fear of being left behind. The consulting company Deloitte helpfully points out: "If you don't adopt automation, your cost base will be dramatically higher than your competitor's." They predict that RPA will "release" people from today's workforce at a rate comparable to the Industrial Revolution.[14]

Most expert discussion of job displacement mentions a time horizon of five to ten years. Many use 2020 or 2025 as the date by which big job shifts are likely to have happened. According to a 2017 survey by Tata Consulting Services, for instance, 80 percent of companies thought AI was essential to their businesses and about half thought of it as transformative

11. See "Artifcial Intelligence in the Real World: The Business Case Takes Shape," *EIU Briefing Paper, Economist.com*, 2016.

12. Estimates from Kate Burgess, "Blue Prism's Rapid Share Price Rise Needs a Reality Check: Robotic Software Group Will Not Make a Profit or Pay a Dividend for Years," *Financial Times*, January 28, 2018.

13. Phil Fersht, "Enterprise Automation and AI Will Reach $10 Billion in 2018 to Engineer OneOffice," *Horses for Sources* (blog), November 4, 2017.

14. Deloitte, *Managing the Digital Workforce*, 2017.

technology. Two-thirds of over eight hundred executives from thirteen global industries thought that digitech was "important" or "highly important" to remaining competitive by 2020. By 2020, half the executives thought the bulk of their digital technology investments would be aimed at transforming their business rather than optimizing existing models.[15]

Taken together—and given the snowball and competition effects that will kick in once the cost-saving job cuts start to materialize—there is a good chance of important disruption by 2020, and a very good chance by 2025. But that's giving the date without the number.

Speed is not the only factor that will turn the Globotics Transformation into the globotics upheaval. Another is the fact that few seem to be preparing for it. There is a very good reason for that. Globots are coming in ways that few expect. This will make it harder for people to prepare and adjust. Indeed, it will probably make it seem like the trend is not a trend at all, but rather a trail of twists and turns. It also means that the pattern of job losses in the last two great transformations will not be very informative today.

WHY GLOBOTS ARE COMING IN WAYS FEW EXPECT

Two days before Christmas 2008, the car assembly plant in Janesville, Wisconsin, closed for good. Then the local car-seat supplier shut down. With thousands suddenly out of work in a town of 60,000, local business suffered. High school students started showing up at school hungry and dirty. Laid-off manufacturing workers retrained for lower-paying service-sector jobs. Thousands of families fell into working poverty. Many entered spirals of despair. The suicide rate doubled.

This outcome—documented so brilliantly in the 2017 book *Janesville: An American Story*, by Amy Goldstein—is how job displacement happened in the Services Transformation. But it is not how jobs

15. Tata Consulting Services, "Getting Smarter by the Day: How AI Is Elevating the Performance of Global Companies: TCS Global Trend Study: Part I," 2017.

will be lost in the Globotics Transformation. Job displacement this time is coming in a new way. The changes will infiltrate our workplaces in ways that are similar to the ways smartphones infiltrated our lives. This requires some explaining.

It Will Happen Like the iPhone "Infiltration"

Just five years ago, the iPhone was a fantastic music player embedded in a mediocre cell phone with a short battery life, a bad camera, and a web browser that wasn't much use (wireless networks were slow and hard to find). Yet one convenience at a time, one cost savings at a time, smartphones infiltrated our lives and our communities.

Smartphones are now our email and messaging center, newspaper, camera, video camera, photo album, dating service, agenda and calendar, travel agent, ticket holder, cash wallet, health tracker, map, yellow pages for finding businesses, web browser, calculator, stock tracker, social media hub, connector of families, source for sports scores, video conferencing facility, ticket agent for movies or whatever, and more. It is even a fairly decent phone (although still has a short battery life).

Smartphones have permeated our lives so thoroughly that many feel naked or even lonely without their phone. And "my phone battery ran out" has become a major excuse for many mistakes. The technology has joined our communities and invited people you don't know to your family dinner table and business meetings. Communities have had to create new rules for these new community members.

But the key point here is that few consciously decided to let this happen. It just happened.

There was no plan; no thinking it through; no government policy. But step by step, smartphones dramatically changed the way we deal with each other, our physical surroundings, and the business and political world. They snuck into our daily routines without us realizing how much they were changing our lives because the advantages seduced us little by little. We can't put our finger on the year that smartphones went from gadgets to life-changers, but after just a few years, we found ourselves asking: "How did we ever get along without them?"

This is how the Globotics Transformation will arrive. Globots will take over professional and white-collar jobs in the same incremental, unreflected way that iPhones invaded our lives. Our companies will bring globots into our workplaces one convenience and one cost-saving at a time. There will never be a "Janesville moment" with which we can date the globotics upheaval. Office and factories will not be shuttered by software robots or telemigrants; the job impact will much harder to detect. It will only be after five to ten years that we'll realize that globots have totally and irrevocably disarranged our workplaces and communities. That's when we'll be asking: "How did we ever get along without them?" In short, the globotics upheaval will be the result from millions of seemingly unrelated choices that we and our companies make.

This steady, accretive nature of digitech's impact on the economy needs a name; I suggest we call it the "iPhone infiltration."

But concretely, how will we know it's happening? The answer lies in easily available statistics—separation and hiring rates.

How It Is Happening in the Information Sector

In a great American tragedy, 5 million workers quit, are fired, or are laid off from their jobs every month. In a great American triumph, about 5 million US workers take up new jobs every month. This fact—which is well known to labor economists—provides a critical insight into how globots will shock the middle class. Telemigrants and white-collar robots will displace professionals and service-sector workers in one of three ways. They may reduce the hiring rate, increase the separation rate, or a bit of both. Consider the example of one sector that has been in the crosshairs of digital technology for a few years already.

The "information industry" is a sector that lives on the gathering, processing, and transmitting of information. It includes jobs like publishing, movies, music, and online services, including Google search. The separation and hiring rates for this sector have been peculiar compared to that of the American nonfarm economy as a whole, as Figure 7.1 shows.

The US economy has boomed since recovering from the global crisis of 2008. Especially since 2012, the overall number of US jobs has risen sharply. The overall rise in jobs, however, was the outcome of a very dynamic process of job creation and job destruction.

The rate of overall new hires per year jumped from 2012 to 2015 and has continued to increase. This rate is shown as the solid black line marked "Hires, Total Nonfarm." The rate of separations (namely, retirements, quits, layoffs, or firings) for the total nonfarm economy has also risen but not by as much (see the dashed black line in the figure). With more hirings than firings in the nonfarm economy as a whole, the number of jobs rose.

Think of this like filling a bathtub with water when the drain is open. If the water flows in (that's the "hires") faster than it flows out (that's the "separations"), then the water level (that's the number of jobs) rises. Put directly, the number of jobs rises when job creation outstrips job destruction. The opposite happened in the "information industry."

The information industry's separations are shown as the dashed grey line in the figure. These have pretty closely followed the total nonfarm

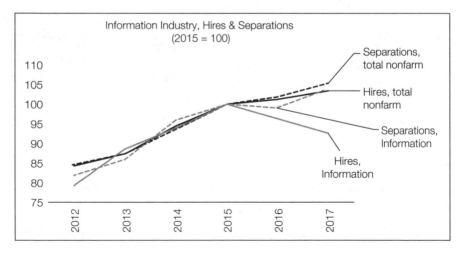

Figure 7.1 Information Industry, Hires and Separations, 2015=100.
SOURCE: Author's elaboration of data published by the BLS.

separations. What is really different is the information sector's hiring (shown as the solid grey line). These dropped off remarkably after 2015. Comparing the two grey lines, we see that the separations outstripped the hires. With more people losing jobs in the sector than were gaining jobs in the sector, the total number of jobs fell. In fact, the information sector lost about 22,000 jobs since January 2015, although that precise number cannot be seen in the figure.

There certainly is a sense of crisis among journalists and other people who used to make their living in this sector, but it was not a *Janesville*-like event. The reduction in jobs was the result of a steady "infiltration" of globots into the newsrooms, editing rooms, and broadcast studios. Many jobs were automated, and others were shifted to freelancers—some of whom were based in low-wage nations.

The next key driver of upheaval—unfairness—has nothing to do with speed, and it is much harder to get a handle on. By their very nature, globots will not play fair. They won't play by the usual rules when competing for human jobs. This matters greatly. Nothing makes people angrier than unfair competition.

The backlashes in the nineteenth and twentieth centuries were greatly accelerated by the fact that the changes were seen as outrageously unfair. And a widespread sense of injustice and outrage were certainly a big part of the 2016 upheavals that produced the election of Donald Trump and Britain's vote to leave the European Union. This is standard.

UNFAIRNESS PUTS THE "RAGE" IN OUTRAGE

The classic example, as we saw, was the Luddite Riots in the early 1800s. Competition from "power looms" led to rapid job displacement, but it wasn't just the job losses that riled up people. Workers saw the power looms as outrageously unjust since they allowed skilled craftsmen with families to look after to be replaced by untrained children who were paid a pittance. This violated long-standing practices. Having seen one set of social norms ignored by the mill owners, the protesting workers felt justified in violating another set of social norms. Things spun out of control. People died.

Hopefully things will not get so dire this time but this example illustrates the importance of focusing on how workers perceive the fairness of their job loss. That is why one simple fact is so important: globots don't play fair.

America's and Europe's middle classes will not welcome the new competition from white-collar robots and telemigrants. The humans will come to view both types of globots as outrageously unfair competitors. Start with the globalization part of globotics.

Unlike the old globalization—when foreign competition meant foreign goods—globotics globalization will involve foreign people who are bringing direct international competition on pay and perks into offices and workplaces. Telemigrants today ask for lower wages and no benefits. Despite this, they find the freelancing pay attractive since they live in low-cost nations and the alternatives in their own countries are often absent.

The other type of globots—white-collar robots—are unfair in similar ways. This is actually one of their selling points. "Imagine a different kind of workforce. A workforce that you can teach countless skills. The more it learns, the more efficient it becomes. It works without ever taking a vacation. It can be small one day or large when your business hits a spike. And it frees up your best people to really be your very best people. Meet the Software Robots—the Digital Workforce."[16] This is the sales pitch on the front page of one of the world's leading providers of white-collar robots.

Another aspect of RPA may dial-up the outrage factor even more. The workers being replaced will be training their robot replacements. Here is how one RPA software company explains it. "WorkFusion automates the time-consuming process of training and selecting machine learning algorithms . . . WorkFusion's Virtual Data Scientist uses historical data and real-time human actions to train models to automate judgment work in a business process, like categorizing and extracting unstructured information." This thing, in other words, is a white-collar robot that figures out what parts of the job can be done by a white-collar robot. And it does it by

16. Blue Prism website, https://www.blueprism.com/, accessed February 4, 2018.

watching what the humans are doing and have done. The program even has a helpful times-up bell. "WorkFusion notifies users when automation can match or exceed the precision level required for a process."[17]

The result, the company claims, lets businesses "reduce manual service effort 50 percent." And then the robots take over routine things. "After training on historical conversations, the Chatbot performs just like a human agent, conversing with customers to achieve context and intent, and executing processes within the back office to fulfil requests." The complex requests are passed on to people—but, like Amelia—the Chatbot learns from how the human resolves the problem, so the people who are not replaced straightaway are, in essence, training the WorkFusion robot to replace them down the road.

Another aspect of globots that will fuel the upheaval is the fact that they are undermining a form of social solidarity—a hidden "welfare system" of sorts. The service sector is where many displaced factory workers have found new jobs. While many of the jobs they obtained were not as good as the ones they lost, they were at least shielded from foreign competition and automation. The Globotics Transformation is changing that.

HOW GLOBOTS UNDERMINE IMPLICIT SOCIAL SOLIDARITY

Most service-sector workers are overpaid in rich nations relative to international standards. To a large extent, this happens because their jobs are sheltered from competition. Economists even have a name for it—the Baumol "cost curse," or the Balassa-Samuelson effect. The basic logic is simple.

Roughly speaking, people get paid according to the value of what they produce. Of course, we can all think of shocking examples of people

17. Workfusion.com blog post, "Intelligent Automation. Digitize Operations with Intelligent Automation for Your Business Processes, with Solutions that Use RPA, Artificial Intelligence, Chatbots and the Crowd", *welcomeai.com*, March 28, 2018.

getting way more or way less than they deserve in terms of value creation, but looking across the hundreds of millions of jobs in our economies, the rough rule is roughly right. The vast international differences in wages and salary are explained by differences in value-creation per hour.

Workers in rich nations generally produce more value per hour than workers in poor nations, but the extra value can come in two ways: productivity or price. In some cases, rich-nation workers are producing more units per hour and the price is not too different. In others, their productivity isn't much higher, but the price is. In many service sectors, it is a matter of prices not productivity. Consider an example.

Germany is a hypercompetitive economy, but not every German is hypercompetitive. One of the most empirically important, but largely unnoticed ways in which the hypercompetitive Germans help the uncompetitive Germans is by paying "too much" for their services. But this does more than transfer income from globalization's winners to its losers.

This strange trick of modern capitalism helps the uncompetitive workers hold their heads high. To put it bluntly, many unskilled workers in rich nations get "overpaid" by international standards, but from a social perspective, they "deserve" their monthly take-home pay. Here's how it works.

German car workers are hypercompetitive. They have very high wages, compared to, say, Polish car workers, but they are cost effective since they produce so much more per hour. That's why German car firms still employ workers in Germany—their superior output-per-hour more than offsets their hourly wages.

Hypercompetitive, by contrast, is not really the right label for German restaurant waiters. German waiters perform about the same tasks and in about the same way as Polish waiters. Yet they get paid far, far more. How can that be in this hyperglobalized world of ours?

The key is that the restaurant sector hasn't been globalized. It is naturally sheltered. German bartenders in Frankfurt are not competing directly with Polish bartenders in Warsaw. People in Frankfurt want to go to restaurants in Frankfurt, and this requires bartenders in Frankfurt—Polish bartenders in Warsaw can't take orders and serve drinks in Frankfurt.

To attract waiters, Frankfurt restaurants have to pay high wages since they have to compete—at least indirectly—with firms in Germany's hypercompetitive sectors like banking, pharmaceuticals, and manufacturing. Of course, waiters don't make as much as bankers, but the high wages of Germany's hypercompetitive workers pull up the whole pay scale. The final piece of this strange trick is that Germans are willing and able to pay high prices in restaurants and bars because they themselves earn a lot.

The Implicit Welfare Payments behind "Overpriced" Services

If you think hard about what is going on here, it is easy to see that this is some sort of tax-and-redistribute scheme that is working though low-skill jobs in sheltered sectors. In essence, the high restaurant prices and wages are one way that Germans who are globally competitive are paying a "tax" which is then distributed directly to boost the earnings of those who are not.

This sharing-and-caring mechanism—which operates across many service sectors—is not exactly Robin Hood robbing from the rich to give to the poor. It is more like a way for the rich to create jobs for Robin's band of merry men so they don't have to rob for a living. It is an indirect way of getting the most competitive citizens to create jobs that allow less competitive citizens to earn a decent living. Moreover, it is all more socially acceptable than charity—on both the givers' and receivers' sides.

In a sense, jobs in a globally uncompetitive service sector have been an important "escape hatch" for workers in rich nations. Trouble is likely to come when globots weld shut this escape hatch via direct wage competition from telemigrants, or direct job destruction by software and hardware robots.

When globots take over this sort of service-sector job, the displaced service-sector workers will start to experience some of the hardships that have been faced by blue-collar workers since the 1980s.

A BED OF DISCONTENT FOR THE UPHEAVAL

At age nineteen, Alfred Perry moved from a declining manufacturing town to a booming high-tech town in North Carolina. He had high hopes and a high school degree in hand: "It was like a rainbow leading to this pot of gold," he said.[18] By twenty-one, he was homeless, having drifted through a sequence of low-paid, dead-end service-sector jobs. If Perry follows the average trajectory of US workers with his skill level, his future could hold some very dark moments.

During the Services Transformation, automation and globalization eliminated good jobs for low education workers. It was the start of what might be called the "wretched ratchet." Manufacturing employment jagged down with each recession and recovered with each recovery, but each time the recovery high was lower than the previous peak. Since 1979, the number of US manufacturing jobs has been on a bumpy, downward slide. Deindustrialization also raised the stakes in terms of education.

Many of the children of the displaced factory workers got university educations to train for service-sector jobs. The workers themselves struggled, but until the last couple of decades, many of them could rely on union membership, experience, and seniority to carry them over till retirement age. And thanks to New Deal policies, many had the means to afford a decent living on their pensions. The shutting of factories threw up another set of issues related to local geography.

Since the industrial revolution, industry has tended to cluster. Much of it was near major metropolitan areas like New York or London, but some factories were situated in smaller towns, especially in the Midwest of the United States, and the Midlands in England. This was a blessing for the local economy when industry was booming, but a curse when manufacturing employment started to decline. A single plant closure could throw the whole community into a tailspin—an outcome that produced the phrase "rust belt."

18. Quoted in Shawn Donnan and Sam Fleming, "America's Middle-Class Meltdown," *Financial Times*, May 11, 2016.

And then there are the "deaths of despair."

Deaths of Despair—Anomie in Action

The lack of good jobs—those with good health insurance policies and other benefits, training, and the expectation of advancement—also made it harder for displaced American manufacturing workers to marry. While most low-skill white Americans born in the 1950s were married by age thirty, the figure dropped to half for those born in 1980.[19] Without the stability of marriage, personal and social instability rose. This class of Americans suffers worse physical and mental health, and more social isolation, obesity, divorce, and suicide.

The mortality rate among US whites aged forty-five to fifty-four with only high school degrees—both men and women—has risen significantly since the late 1990s. "Half a million people are dead who should not be dead," writes Nobel Prize winner economist Angus Deaton with his co-author Anne Case, a professor of economics at Princeton University. They call these "deaths of despair," and they find that they have been rising across the US at every level of urbanization.

The proximate causes of the higher death rates are clear—drugs, alcohol, and suicide—but Case and Deaton view them as all the same: "In a sense, they are all suicide—either carried out quickly (for example, with a gun) or slowly, with drugs and alcohol." Case and Deaton believe that the higher jobless rates, reduced marriage rates, and worse physical and mental health of Americans caused the higher death rates indirectly. It did this by kicking out the social and economic supports that used to help people get through hard times.

The Case-Deaton view echoes Durkheim's theory of "anomic suicide." Anomie—namely disconnection from society, a feeling of not belonging, and weakened social cohesion—can make people feel so estranged that

---

19. Anne Case and Angus Deaton, "Mortality and Morbidity in the Twenty-First Century," BPEA article, March 23, 2017.

they commit suicide.[20] Durkheim, writing in 1897, suggested that anomie is especially prevalent during times marked by socioeconomic and political convulsion that lead to rapid and extreme changes in people's communities and everyday lives.[21]

Case and Deaton recast the theory in modern terms. The deaths are, as they put it, the outcome of "cumulative disadvantage."

Cumulative Disadvantage

Case and Deaton conceive of people as being handed various "burdens" throughout their lives. The heavier the burden, and the longer it has to be borne, the harder things get. And starting from the 1970s, the burdens piled on for this group.

Life didn't turn out as they were raised to believe it would. When the American Dream became the American Illusion, a sense of hopelessness crept in. People turned to overeating and alcohol or drug abuse. They no longer turn to standard social organizations like traditional churches, marriage, and family; without these stabilizing social structures, things could, and often did, spin out of control. As Deaton puts it: "We are trying to say that low income and low job opportunities, after a long period of time, tears at the social fabric. It's the social fabric that keeps you from killing yourself."[22]

This trend is mostly an American phenomenon since, in the US, social market capitalism became a lot more market and a lot less social than it was before President Reagan started to undo the New Deal. Low-education, middle-aged Europeans and Japanese have suffered the same effects of automation and globalization, but they were supported by cohesive social

20. See excerpt from Robert Alun Jones, *Emile Durkheim: An Introduction to Four Major Works* (Beverly Hills, CA: Sage, 1986), 82–114.

21. Émile Durkheim, *Suicide: A Study in Sociology* (1897; repr., New York: The Free Press, 1951).

22. Alana Semuels, "Is Economic Despair What's Killing Middle-Aged White Americans?" *The Atlantic*, March 23, 2017.

fabrics and government-sponsored safety nets. Their governments routinely provide financial support, healthcare, child support, and pensions that relieve individuals of much of the cumulative disadvantage.

The classes of workers that had made up Roosevelt's forgotten men and women were being forgotten anew. Rather than easing the painful impact of automation on industrial workers, the US political system made things worse. A half century after the New Deal, government policy was again driven largely by the money and political power of the "one percent"—just as it had been in nineteenth-century Britain.

Taxes for the rich were lightened as safety net services for struggling Americans were cut. In one particularly important policy change, individual Americans were limited to five years of welfare benefits for their whole life. Those who have exhausted this limit have nothing to fall back on. As part of this trend, anti-union laws were passed at the state and federal levels with President Ronald Reagan a notable champion of this policy. Labor market regulations were relaxed, union membership declined, and many aspects of the social safety net were weakened in the name of pro-market, business-friendly reforms.

FROM UPHEAVAL TO BACKLASH

The Globotics Transformation is playing with fire around a powder keg of discontent—especially in the US where the safety net is set far too low to be of help to many Americans who have borne the brunt of the disruption that the Services Transformation injected into the system since 1973.

One of the great economic historians of our times, Barry Eichengreen of the University of California—Berkeley, dissected the 2016 backlash by putting it into historical context in his 2017 book, *The Populist Temptation: Economic Grievance and Political Reaction in the Modern Era*. Drawing on examples from the 1800s onward, he sums it up this way: "Populism is activated by the combination of economic insecurity, threats to national identity, and an unresponsive political system." The resulting populist backlashes are often damaging and destructive.

"Populism arrays the people against the intelligentsia, natives against foreigners, and dominant ethnic, religious, and racial groups against minorities."

The economic insecurity, hardship, and despair created by the disruptive duo's impact on the US and European economies from 1973 had political consequences that we saw in 2016. The economic insecurity and perceived threats to national identity that are coming with the Globotics Transformation seem destined to lead to further backlash since the political systems in the US and Europe are unresponsive to the challenges so far. The governments are either unaware of how fast the changes are coming or living in denial about their implications for middle-class prosperity.

The factors that are turning the Globotics Transformation into the globotics upheal are clear to see and already in operation. If history is a guide, the next step will be some form of backlash, and possibly another wave of populism.

It has happened before.

8

The Globotics Backlash and Shelterism

On the morning of November 30, 1999, Seattle police woke up to find that the "antiglobalization movement" had started. It was just that quick. Before November 30, there had been antiglobalization "moments"; on November 30 the moments became a movement.

The night before, ten thousand protestors surrounded the Paramount Theater and Convention Center where the pro-globalization World Trade Organization (WTO) was supposed to have its opening ceremony the next morning. The Seattle police were unaware and unprepared. The mass civil disobedience won the day and the opening ceremony was canceled. But the day wasn't over.

In another part of the city, twenty-five thousand labour unionists started a peaceful march. When they reached downtown, the combination of environmentalists and unionists stretched police capacities. Black-hooded anarchists seize the opportunity to smash windows and burn cars. By midday, Seattle was a mess.

The National Guard and US military units were called in and an overnight curfew was enforced. Protesters were teargassed and beaten with batons. Five hundred people were arrested. The city suffered millions of dollars in damages in what came to be known as the "Battle in Seattle."

And then it spread globally.

The antiglobalization movement burst onto the international stage at a pace that astonished and amazed—blindsiding authorities in many nations. The years 2000 and 2001 witnessed massive, antiglobalization protests in Washington DC, Prague, Nice, and Gothenburg in Sweden. Things turned violent in Sweden. Overwhelmed by the number of protesters, the Swedish police used batons, horses, dogs, and eventually guns to control the crowd. Police shot three protesters. Radicals responded with bricks and Molotov cocktails. But things got even more radical at the next G8 summit in Genoa.

Three hundred thousand demonstrators gathered outside a meeting of the G8 heads of state in Genoa, Italy to face off tens of thousands of police. In preparation, a thirteen-foot fence was set up to protect the heads of state. Train and plane services into Genoa were suspended, highway exits were blocked, and a special watchlist was established to deny entry into Italy of known anarchists. Despite the preparations, violence erupted. A twenty-three-year-old protestor, Carlo Giuliani, was shot dead by police. Hundreds were injured. Hundreds were arrested. The city center looked like a war zone.

There are important lessons here for how the globotics upheaval could turn into a violent globotics backlash.

BACKLASH BEDFELLOWS—FUSING THE FURIES

A peaceful protest by nature lovers turned into the "Battle in Seattle" because of the unlikely fusion of unlikely bedfellows—environmentalists, labor unionists, and anarchists. A *Washington Post* journalist wrote at the time: "What's really surprising is that the people who don't like free trade—the Pat Buchanans and Ross Perots, the unions, the environmentalists, the freaks, the randomly angry people—were somehow able to stand one another's presence long enough to organize a massive protest."[1]

---

1. Joel Achenbach "Purple Haze All Over WTO", *Washington Post*, December 1, 1999.

Globalization in the 1990s had made different groups furious for different reasons and these differences had long kept them from cooperating. At Seattle, the furies fused. If the globotics upheaval does flare up into a violent backlash, my guess is it will involve a similar fusion.

For decades, millions of blue-collar workers have been competing with Chinese manufacturing abroad and industrial robots at home. Neither competition has been going well. Automation and globalization damaged these workers' financial prospects and have thrown their communities into disarray. These blue-collar workers will soon have company.

Various experts predict that globots will displace millions, tens of millions, or hundreds of millions of service-sector and professional workers. If it turns out to be "only" millions and the changes are spread out over many years, the globotics upheaval will stay contained. If it is hundreds of millions and it happens in a few years, the results could be revolutionary in the bad sense of the word. Quite simply, the globotics upheaval's disruption of service-sector and professional jobs will be like tossing a lighted cigarette into a firework factory.

This combination of blue-collar and white-collar voters will be an unstable mixture. It is the type of combination that has in the past exploded. In the early twentieth century, lingering economic difficulties induced Europeans to long for authority, justice, and economic security. This led them to embrace extreme solutions (fascism or communism). Things probably won't go that far, but the feelings are not so different today, especially in the US.

A Base of Anger—the 2016 Backlash that Gave Nothing Back

Patti Stroud knows all about the disruptive impact of the globots that triggered American deindustrialization. For a quarter-century, her husband had a good job at a steel mill in Pennsylvania. It was closed just weeks before the 2016 US presidential election.

The fifty-six year old, who cleans houses for a living, voted for Trump because he promised a break from the past. "I thought we needed a big change, and boy, did we get it," she said in a March 2018 interview with the *New York Times*.[2] But it was not the change she was hoping for.

Trump and Brexit voters were angry in 2016—frustrated with mainstream politicians who had failed to stop the disruptions of their communities, the loss of good jobs, and the relentless undermining of the hope that things would get better. For far too long, they bore the brunt of the tech-trade team's disruptive influences. They were the bearers of far too much of the "pain" part of the gain-pain package that automation and globalization has been delivering to the working class from 1973. Voting to leave the EU and electing an unruly outsider as US president were ways of saying "enough is enough."

But in fact, the 2016 backlash has given very little back. The 2016 populists politicians offered illusion-based solutions—like a border wall, or leaving the EU—to reality-based problems—like deindustrialization, or stagnate wages. These voters are still struggling financially. Neither Trump nor Brexit have improved things for them materially. The economic calamity continues—especially in the US.

The loss of manufacturing jobs has fundamentally damaged the lifetime prospects of many Americans. There is only the thinnest chance that a fifty-year-old worker displaced from manufacturing will find a job that pays as well or provides as much income security. This reality created a sense of hopelessness, a sense that a good new job is not on the way, that wages for the jobs on offer will never rise, and that fractured communities will never coalesce again. And the lack of hope is teamed with poor outcomes.

The US numbers are sobering. Forty million live in poverty and half of those earn less than half the poverty-level income.[3] A quarter of US children live in poverty. The country has the highest rate of obesity in the

2. Quotes from Trip Gabriel, "House Race in Pennsylvania May Turn on Trump Voters' Regrets," *New York Times*, March 2, 2018.

3. Numbers from "Income and Poverty in the United States: 2016", by J. Semega, K. Fotenot, and M. Kollar, US Census Bureau, September 2017, and Yale's Environmental Performance Index, http://archive.epi.yale.edu/epi/issue-ranking/water-and-sanitation, and https://www.vox.com/2015/4/7/8364263/us-europe-mass-incarceration

developed world, and it ranks below Lebanon in terms of access to water and sanitation. The share of the US population in jail is the highest in the world; it is five times higher than the average among rich nations.

US men in particular have just been giving up in record numbers—especially those with only high school educations. The share of prime-age males (twenty-five to fifty-five) in work or looking for work has fallen steadily since the 1970s, with the trend noticeably more marked for those with a high school education or less. In 1974, the participation rate was 92 percent; in 2015, it was about 82 percent. The rate for those with college degrees fell as well, but only from 97 percent to 94 percent.

And the future looks no brighter for the people hardest hit by deindustrialization. US economic mobility has dropped steadily since the 1970s. Eighty percent of Americans born in 1970 into a household with an average income would achieve higher incomes than their parents. Kids born in an average household in 1980 have only a fifty-fifty chance of doing better than their parents economically. And in the hard-hit Midwestern states, the situation is worse. There is a better-than-even chance that the kids of average parents will slip down the economic ladder.

Almost half of middle-aged Americans have too little money saved for a comfortable retirement. A recent survey showed that 40 percent could not come up with $400 to cover an emergency without borrowing or selling something. One out of four had to deny themselves some form of healthcare since they couldn't afford it.

In the US, healthcare is still getting more expensive, the debt-financed tax cuts went mostly to the richest Americans, and nothing has been done to help displaced workers adjust to twenty-first-century economic realities. In Britain, public services continue to deteriorate and almost nothing has been done to reinforce the adjustment policies aimed at assisting workers affected by deindustrialization. In both countries, many voters still feel their communities are under threat culturally as well as economically and the result has been growing anti-foreigner feelings.

History is littered with examples of discontent that led to nothing more than a disorganized mass of angry and frustrated people. But it doesn't always end that way. Sometimes a group of individuals turns into an individual group and the result can shift history. The process is messy and not well understood—as is true of all complex social happenings.

COULD THE BACKLASH PRODUCE VIOLENT PROTESTS?

"I think the great majority of people who have joined this movement started off with a vague sense that something was wrong and not necessarily being able to put their finger on what it was," said George Monbiot, a columnist for the British newspaper *The Guardian*.[4] He was talking about the process that turned many antiglobalization "moments" in the 1990s into a giant antiglobalization movement, but the quote fits today's mood. In many parts of the US, Europe, and other high-income nations, there is a generalized feeling of vulnerability, exploitation, and injustice—but no clear sense as to who is to blame.

A "vague sense that something is wrong" does not produce mass demonstrations and street violence. Movements need targets to focus the anger. The target of the antiglobalization movement turned out to be multinational corporations but the target emerged organically.

Monbiot explains that various activists were "having a sense that power was being removed from their hands, then gradually becoming more informed, often in very specific areas." At first there seemed few connections. There were "some people who are very concerned about farming, those who are very interested in the environment, or labour standards, or privatisation of public services, or Third World debt." The thing that connected the dots was large corporations: "These interests tie together and the place they all meet is this issue of corporate power," wrote Monbiot.

Multinationals, especially the big tech companies, may turn out to be the target of the globotics backlash if it does go global and does go to the street.

4. Quotes from Mike Bygrave, "Where Did All the Protesters Go?" *The Observer*, July 14, 2002.

Who Might the Backlash Target?

Big tech companies like Facebook, Amazon and Google were just starting to get roughed up in the "playground" of public opinion when this book went to press. In early 2018, Mark Zuckerberg, CEO of Facebook was called to testify before the US Congress and EU Parliament about a scandal involving the misuse of users' data.

These guys make perfect targets for populist backlashers. They are fabulously rich for one. Zuckerberg's estimated weatlth is over $70 billion; for comparison, the US Marine Corps' annual budget is only $27 billion. On top of that they are well-known to the general public, and some of their companies are involved in the automation of white-collar jobs and online freelancing. Another aspect that will make them targets-for-opportunists is a vague sense that these men (and they all are men) and their corporations are exploiting, for personal profit, the most human of tendencies—the need for sharing with others.

One line of attack so far has stressed the manipulative nature of the services. "We talk about addiction and we tend to think, 'Oh, this is just happening by accident'" said Tristan Harris, who was the CEO and co-founder of a startup Google bought in 2011 before becoming a design ethicist and product philosopher at Google. "The truth is that this is happening by design. There's a whole bunch of techniques that are deliberately used to keep kids hooked."[5]

Harris mainly wants to raise awareness of the problem with an anti-digitech addiction campaign aimed at fifty-five thousand US schools. Others are more accusatory. "The largest supercomputers in the world are inside of two companies—Google and Facebook," notes Chamath Palihapitiya, a venture capitalist and early Facebook employee. "The companies are "pointing them at people's brains, at children." The result, he argues, is "ripping apart the social fabric of how society works." Promoters of this line of outrage seem to be viewing things in a way that

5. Quoted in David Mogan, "Truth About Tech Campaign Takes on Tech Addiction," *CBSNews. com*, February 5, 2018.

is not too far from the zeal shown by the "Temperance Movement" of the 1910s that led to a constitutional amendment against alcohol in 1920.

The allegations of evil-doing and greed could provide the focal point for protest. The former prime minister of Belgium, Guy Verhofstadt, put the point directly to Facebook CEO Zuckerberg when he was testifying before the EU Parliament. "You have to ask yourself how you will be remembered. As one of the three big internet giants together with Steve Jobs and Bill Gates who have enriched our world and societies, or on the other hand, as the genius that created a digital monster that is destroying our democracies and our societies?"

Of course, this sort of allegation is a long way from the job-displacing effects of globots, but as we saw in the antiglobalization movement, the targets of the backlash often find themselves in a crossfire from people with diverse grievances. Yet another source of what could become a mighty pushback takes an even deeper bite at the big tech companies by focusing directly on their goldmines—their data.

RADICAL MARKETS AND WHO CONTROLS OUR DATA

Two Chicago University scholars, Eric Posner and Glen Weyl, published a book in 2018 that points out that no one thought through the "data economy" before it happened. Their book, *Radical Markets: Uprooting Capitalism and Democracy for a Just Society*, argues that the data-based economy unknowingly developed without any systematic thought as to the consequences. Its design was driven by greed and human curiosity. The result, they argue, is inefficient and unproductive as well as being unfair, so radical solutions are needed.

The solutions they propose could easily be part of a backlash against globots.

The authors point out that today data is governed by the "data-as-capital" view. Once we give our data to these companies, it is theirs to keep. They get to use it as much as they like and however they like. It is like you have donated a book to a public library and the librarian gets to decide what to do with it. They suggest a radically different solution, what they call "data-as-labor." Data in this view is generated by users and thus the data belongs to the users. If the big tech companies want to use it, they have to pay the users. Just imagine the radical implications of that simple switch in data ownership.

Under the data-as-labor presumption, digitech firms would have to pay people for the data they create. Suppose the parliaments of all the advanced economies passed laws that forced Facebook (to take an example) to pay each of its users $100 per year for the right to use their data. This would generate a wider distribution of income and cultivate "digital dignity."

The EU's digital law, the General Data Protection Regulation, is a step in this direction. It protects and empowers EU citizens when it comes to their data privacy and the ownership of their data. It is already reshaping the way organizations interact with online users.

Financial Times columnist John Thornhill puts it this way: "We consumers should wise up to our role as digital workers and—in Marxist terminology—develop 'class consciousness.'" He suggests the formation of "data labor unions" that could fight for our collective rights. Somewhat tongue-in-cheek, he predicts that we'll know this is getting serious when people start "digitally picketing social media groups." He even has a quip for the placards: "No posts without pay!"

This "radical markets" solution is indeed radical. And it is easy to see how it might find allies among the Teamsters, lawyers, and office workers who will lose their jobs to white-collar robots and telemigrants.

If the globotics upheaval and backlash turn into something big and violent, we will need to see a process like the one that brought the antiglobalization movement into existence. But what governs such processes? How do a group of individuals turn into an individual group? The answers, imprecise as they are, come from sociology.

From Individual to Collective Action

One pioneering sociologist, Émile Durkheim, viewed people has having two levels of existence—two personalities, so to speak. At one level—the level that is apparent most of the time—individuals care about themselves

and their loved ones. At the other level, individuals submerge their indi-
viduality. They act as if their interests and the interests of the group were
the same. They follow the group's actions and obey the group's direction
even when doing so harms their personal interests.

These two levels constantly coexist within each of us, according to
Durkheim, but they don't operate at the same time. This can generate
seemingly paradoxical behavior. A young man can, for instance, cheat
on his taxes to save a bit of cash. But the same man can, in different
circumstances, be willing to die for the country that he was cheating.

A critical question is: what triggers the shift between levels? What
switches people from operating on the individual level to operating on the
group level? Jonathan Haidt, author of the influential book, *The Righteous
Mind: Why Good People Are Divided by Politics and Religion*, has a name
for this. He calls it the "hive switch." Flip the hive switch and the self shuts
down and the groupish instinct takes over, making people feel like they
are part of something greater than themselves.

Haidt argues that the flipping of just such a hive switch was a critical
aspect of the 2016 backlash—especially the election of Donald Trump. The
authoritarian aspects of Trump appealed to many Americans who were
reacting as members of communities under threat—not just individuals
facing economic difficulties. As Haidt wrote, many Americans "perceive
that the moral order is falling apart, the country is losing its coherence and
cohesiveness, diversity is rising, and our leadership seems to be suspect."
In such situations, a goodly share of the population instinctively reaches
for autocratic solutions. "It's as though a button is pushed on their fore-
head that says: 'in case of moral threat, lock down the borders, kick out
those who are different, and punish those who are morally deviant.' "[6]

The globotics transformation won't be as obvious as the deindustrialization that has plagued America's middle class for decades. The office
automation will not force whole office buildings to shut down. Globots
will be slipped in one by one. The transformation will look more like the

---

6. Jonathan Haidt, "The Key to Trump is Stenner's Authoritarianism", *The Righteous Mind*
(blog), January 6, 2016.

iPhone infiltration than the Janesville factory closings. This will make it hard for people to identify the trend. But there will always be populists willing to point fingers and make exaggerated claims against their targets as a way of gaining power.

In the US, one such populist is already a declared candidate for the 2020 presidental election. His name is Andrew Yang, the presidential wannabe we met in the Introduction. On his campaign site, Yang put it in stark terms.[7] "Good jobs are disappearing. New technologies like robots and AI are great for business, but will quickly displace millions of American workers. In the next twelve years, a third of all American workers are at risk of permanently losing their jobs, a crisis far worse than the Great Depression."

He seamlessly weaves the woes of blue-collar workers with those of white-collar workers hit more recently by globots. A "massive employment crisis is already underway. . . . Artificial intelligence, robotics, and software are about to replace millions of workers. This is no longer speculative—it is already happening." There is, he asserts, a very real threat facing tens of millions of Americans, everyone from truck drivers and lawyers to call center workers and accountants.

He predicts a violent backlash. "All you need is self-driving cars to destabilize society. We're going to have a million truck drivers out of work. That one innovation will be enough to create riots in the streets."

Yang embraces the standard stance of populist-as-outsider. Someone who can stand up for the people (who are pure) and against the elite (who are corrupt). "I'm not a career politician—I'm an entrepreneur who understands technology and the job market, and I know things are going to get much, much worse than the establishment is willing to admit."

His solutions, which he writes about in his 2018 book, *The War on Normal People: The Truth about America's Disappearing Jobs and Why Universal Basic Income Is Our Future*, are not revolutionary. This isn't a new "ism" like fascism or communism. But we are living through a volatile period, and things could easily get out of hand. Yang puts it starkly: "We

7. "Andrew Yang for President" website, www.yang2020.com.

have two options. We can stay the course, and let millions of hardworking Americans fall into unemployment and despair. Or we can face the challenge together, and create a society in which humanity is valued as much as the market."

The emergence of populists like Yang is quite predictable given the disruptive nature of globotics. His themes will surely get more mainstream as the 2020 presidential election approaches. But the timing of any such blowup is impossible to pin down.

Social psychologists tell us that violent protest is best understood as an irrational thing—an emotional thing that is often triggered by a sense of injustice.[8] A classic example of this came a couple years after the Battle in Seattle. In 1992, four white Los Angeles police officers who had beaten a black motorist, Rodney King, after a high-speed chase were acquitted of assault. A video of the original incident convinced many in Los Angeles that this was a clear-cut case of police brutality, so the acquittal triggered emotional outrage. The result was five days of violent backlash. A dusk to dawn curfew was declared. The National Guard was called in. Over fifty people died, thousands were injured, and over a thousand buildings were partly or completely destroyed. Globots displacing workers won't trigger this sort of sudden rioting, but it illustrates how emotional and violent backlashes can be.

There is nothing smooth or predictable about the process that puts the "rage" in outrage, as some recent social science research shows.

Shared Unfairness Puts the "Rage" in Outrage

In a fascinating study of the "dynamics of outrage," Nobel Prize winner Daniel Kahneman and colleagues ran experiments of "mock" trial juries involving over 3,000 people and 500 juries. The idea was to see whether people talking among themselves about unfairness led the group as a whole

---

8. Samantha Reis and Brian Martin, "Psychological Dynamics of Outrage against Injustice," *The Canadian Journal of Peace and Conflict Studies*, 2008.

to be more or less aggressive in terms of punishment than the individual jurors before the discussion. In other words, does the group (the jury) act more radically than the group of individuals (the jurists individually).

Six-person juries were presented with evidence of a mock personal-injury case, and then asked—individually—to say how much they think the guilty party should pay to the victim. Then the six individuals talked over the case among themselves to decide the appropriate punishment.[9] The findings are useful in understanding why it is so hard to predict when social upheavals turn violent.

When the typical juror felt the "mock" crime was truly outrageous, the jury as a group got extra harsh. In other words, the fact that the crime was outrageous made the individuals as a group act in more extreme ways than the average of the individuals deciding on their own. "Mob mentality" would be an unscientific phrase for it. Outrageous things seem more outrageous when you share your sense of outrage with others. And a similar thing happened in the opposite direction. When it came to mock crimes that seemed trivial or technocratic, the group as a whole acted more leniently after they deliberated together.

The key point here is that this sort of group dynamics makes social outrage into a highly unstable, highly unpredictable thing. Cass Sunstein wrote a recent article discussing the key role that injustice played in the rapid spread of the #MeToo movement. He stressed the point that the reaction outrage causes can depend upon unexpected dynamics. "With small variations in starting points, and inertia . . . [o]utrage may fizzle or grow."[10]

Another key point—and one that reinforces the notion that the globotics backlash will involve a fusing of white-collar and blue-collar furies—is that outrage usually springs from a bed of long-lived discontent. Economic hardship and extremism are long-time, historical companions.

9. Cass R. Sunstein, David Schkade, and Daniel Kahneman, "Deliberating about Dollars: The Severity Shift," Law & Economics Working Papers No. 95, 2000.

10. Cass Sunstein, "Growing Outrage," in *Behavioural Public Policy*, 2018 (in press).

Economic historians have found that severe and prolonged economic shocks have political consequences. Drawing lessons from the political history of twenty countries going back to 1870, a team of economic historians found that democracies tend to take a turn towards far-right politics following severe economic shocks (specifically, financial crises).[11] Far-right vote shares rose, on average, by about a third in the five years after the shock. On top of this, and perhaps feeding the trend, governing got harder. Parliamentary majorities shrank and the number of parties in parliaments rose. As a result, decisive political action became more difficult just as it was most needed.

The impacts of the economic shocks were not limited to elections. Lingering economic shocks were associated with backlashes that spilled out on to the streets. General strikes were a third more likely, riots were twice as likely, and antigovernment protests were three times more likely after major economic shocks.

It is not just the size of the shock that matters. Another set of economic historians, led by Berkeley professor Barry Eichengreen and Oxford professor Kevin O'Rourke, found that long-lasting economic troubles were particularly associated with a rise in the share of votes won by right-wing parties.[12] Support for far-right, populist parties grew the most when economic hardship was allowed to persist for years—as it has been allowed to do in America in recent decades.

So will the globotics backlash turn to extremes and violence? This is not a question that can be answered with certainty. There is nothing sure about a violent backlash, but it is a possibility we should think about. What is sure is that there will be at least a milder form of backlash that I call "shelterism."

Shelterism means the sorts of policies people want when they are not bent on stopping progress, but still want some "shelter from the storm."

---

11. Manuel Funke, Moritz Schularick, and Christoph Trebesch, "The Political Aftermath of Financial Crises: Going to Extremes," CEPR policy portal, *VoxEU.org*, November 21, 2015.

12. Alan de Bromhead, Barry Eichengreen and Kevin O'Rourke, "Right-wing Political Extremism in the Great Depression," *VoxEU.org*, February 27, 2012.

Indeed, it's already started. Politically powerful groups that are threatened by digitech are calling for and getting regulatory shelter that slows or reverses the changes.

BEST BET BACKLASH: SHELTERISM

Eight thousand drivers of London's iconic black cabs brought central London to a standstill with a drive-slow protest in February 2016. They were protesting against digitech—or, more precisely, against Uber. Uber had taken millions of rides that would have gone to black cabs. In objecting to this, Steve McNamara, head of the drivers' association, didn't focus on the economic competition—he focused on the unfair and unsafe bits. "Since it first came onto our streets, Uber has broken the law, exploited its drivers and refused to take responsibility for the safety of passengers."[13]

Uber is neither a white-collar robot nor a telemigrant, but it turned taxis from a sheltered sector to an open sector—just as globots are doing in many service sectors. And, like power looms in northern England in 1811, the technology seemed outrageously unfair. Skilled workers saw their occupations suddenly opened to competition from less qualified, less regulated workers.

The go-slow protest is a classic example of how workers will react when their livelihoods and communities are threatened by technology (or globalization), especially when the changes are viewed as unjust. Drivers wanted some shelter from the shock. And in fall 2017, they got it.

Urged on by a left-leaning mayor, London Transport removed Uber's license, saying the company was not "fit and proper" to operate. Passenger safety was a big issue, and London Transport noted that Uber had failed to inform authorities about crimes committed by drivers, including one case of sexual assault. But the safety-based rationale wasn't the only, or perhaps even the main, motivation for the opposition to Uber. Cabbies were

13. Quote from Sarah Butler and Gwyn Topham, "Uber Stripped of London Licence Due to Lack of Corporate Responsibility," *The Guardian*, September 23, 2017.

bearing most of the cost of the new technology and the ban was one way of sharing the pains and slowing the inevitable integration of Uber-like technology into the industry. The ruling stands despite a pushback against the backlash—40,000 Uber drivers and 850,000 of their riders signed an online petition requesting a reversal. The ban has spread to other British cities (Uber is challenging these in court).

Health, safety, environmental, and—above all—privacy regulations are the obvious means of slowing the disruption duo's impact on livelihoods. This will be easier in sectors that are already heavily regulated—like banks and motor vehicles—since imposing rules on, say, robo-reporters would require a whole new regulatory infrastructure with surveillance and enforcement mechanisms. Setting these up is possible, but will take much more time than denying a license to Uber.

One good example of shelterism in action is the way American truck drivers are agitating for regulatory shelter from self-driving vehicles.

Regulatory Shelterism

A globot killed Joshua Brown, or so some would claim. In May 2016, his Tesla collided with a truck. He died instantly. Despite safety issues raised by this and other accidents, US states are pushing forward laws that will hasten the progress of vehicles driven by software robots. In December 2016, for example, Michigan allowed the testing and use of self-driving cars on public roads, including ride-sharing and truck platoons (where a few robot-driven trucks follow each other closely). The Michigan law doesn't require a human to be in the vehicle. This has truckers worried, and their labor union is doing something about it.

The truckers' union, the International Brotherhood of Teamsters, is over a century old—having been formed when a teamster was someone who drove a team of horses. James Hoffa, the union boss, said, "I'm concerned about highway safety. I am concerned about jobs. I am concerned we are moving too fast in a very, very strategic area that we have to make sure we are doing it right because lives are involved. . . . It is vital that Congress ensure that any new technology is used to make transportation safer and more effective." Hoffa claimed, big business is running this regulation. Big business's goal, he asserted, was to "get drivers out of the seat and make money... If a guy makes $100,000 for driving a truck where is he going to get a job like that?" But he doesn't want to be viewed as a modern Luddite, as he adds: "Obviously we can't stop progress."[14] The sentiment is finding a voice. Two New York lobby groups, Upstate Transportation Association and Independent Drivers Guild, pressed for bans on autonomous vehicles to avoid losing thousands of transportation jobs.

Labor came out OK in this battle of big business and big labor. The US House of Representatives passed a bill on self-driving vehicles that was generally pro-automation with one notable exception—trucks. The legislation, which still hadn't made it into law when this book went to print, grants nationwide permission for up to 100,000 vehicles to be tested without safety approval, but explicitly excludes commercial trucks.[15]

Since the Joshua Brown accident involved a robot-driven car and a human-driven truck, it is easy to believe that a law which allows robots into cars but not trucks is not only about safety. It surely matters that a rapid shift to robot-driven vehicles could displace over four million workers in the US, with taxi, bus, and truck drivers leading the ranks.[16] And it may have helped that the Teamsters have many members in Midwestern states, and they will be critical in the 2020 US elections.

Motives were clearer in the January 2018 talks between US shipping giant UPS and its 260,000 union workers in North America. The Teamsters are asking UPS to commit to replacing no drivers with drones or self-driving trucks. None of this is explicitly anti-technology. The thrust seems to be

---

14. See David Shepardson's "Union Cheers as Trucks kept out of U.S. Self-Driving Legislation," *Reuters.com*, July 29, 2017.

15. Keith Laing, "Senators Drop Trucks from Self-Driving Bill," *Detroit News*, September 28, 2017. The House version of the bill had passed by the time this book went to press; the Senate version was pending; Chris Teale, "US Senate Considers 'Different Possibilities' to Pass AV START Act," *SmartCitiesDive.com*, June 14, 2018.

16. "Stick Shift: Autonomous Vehicles, Driving Jobs, and the Future of Work", Center for Global Policy Solution, March 2017.

to protect particular groups. As the Amercian TV presenter Malcolm Gladwell put it: "I wonder if we aren't at the beginning of an extended period of backlash in this country . . . where in the face of overwhelming amounts of change in a very small time what people basically say is, 'Let's stop. Enough.' "[17]

AI-driven vehicles are perhaps the most obvious target for shelterism, but the trend is spreading. The US Congress is taking the first steps towards broader regulation. In December 2017, senators and congressmen introduced a bill to set up a federal advisory committee that would evaluate the broader impact of AI on the US economy and society. "It's time to get proactive on artificial intelligence," said Representitive John Delaney. "Big disruptions also create new policy needs and we should start working now so that AI is harnessed in a way that society benefits, that businesses benefit and that workers benefit."

These politicians have good reason to be proactive. Recent opinion polls show that US voters support regulatory pushback against the globots. In 2017, the Pew Research Center surveyed American attitudes toward the globotics transformation—or, as they put it, a world where "robots and computers are able to do most of the jobs that are done by humans today." Over three-quarters of the respondents thought the scenario of robots and computers taking over many jobs currently done by humans was realistic.

The poll also showed firm support for shelterism.[18] Almost six in ten Americans thought that the government should impose limits on how many jobs businesses can replace with machines. Only 40 percent felt businesses were justified in replacing humans with machines simply because the robots cost less. More than eight in ten said they favored limiting machines to "performing primarily those jobs that are dangerous or unhealthy for humans."

17. Quoted in "Anxiety about Automation and Jobs: Will We See Anti-Tech Laws?" James Pethokoukis, *www.AEI.org* (blog).

18. Quotes from Luke Muelhauswer, "What Should We Learn from Past AI Forecasts?," Open Philanthropy Project, September 2016.

Taken together, these opinions suggest that the US electorate is ready for shelterism. Voters are primed for policies that slow down the job displacement—at least in the abstract. Another example of an existing policy that slows job displacement is the law that the European Union adopted to deal with real migration, not telemigration.

Social Dumping in Europe

"This is an important step to create a social Europe that protects workers and makes sure there is fair competition," said Agnes Jongerius, a Dutch Labor Party member of the European Parliament. She was reacting to a reform of something that is akin to telemigration, but without the "tele," namely temporary work done by workers from one EU nation in another.

The EU is, in principle, a single market when it comes to labour. That means that companies from one EU member can bring their own workers when doing projects in other EU nations. For example, a Polish construction company can use Polish workers on German building sites—paying them Polish wages and paying Polish social charges (the European equivalent of US payroll taxes like Social Security). The workers themselves continue paying Polish taxes even though they are working in Germany. There are lessons here for the likely reaction by American and European workers as telemigrating gains in popularity.

This practice of using lower-paid foreign workers led to outrage on the part of German workers. There is even a name for this unfair competition—"social dumping"—which means undermining work conditions in the host country due to increased competition from workers with laxer workplace regulations, wages, and/or taxes. This label is a very conscious analogy with dumping as it is used by international trade lawyers. When trade lawyers say "dumping", they mean exporting goods at prices that are below production costs. The "social" is added to indicate that the goods are made in countries with weak social protection. This practice led to a backlash and the imposition of a form of regulatory shelterism called the Posted Workers Directive.

The way this shelterism arose provides a good illustration of the unpredictable dynamics of social outrage and upheaval. While free migration within the EU has been a reality since the 1990s, concerns about social dumping were fairly marginal for years. But then the movement picked up momentum. The large wage and tax gaps in Europe, plus the post-2008 growth slowdown, led to rising numbers of posted workers and a political backlash. As the EU president Jean-Claude Juncker stated in 2014, "in our Union, the same work at the same place should be remunerated in the same manner."[19] The reform that Junker was referring to—the Posted Worker Directive—limits the duration of posted-worker jobs to twelve months. After that, the worker has to be paid and employed according to local laws.

The rise of telemigration will spark a similar reaction. Local workers will surely come to view telemigration as "social dumping"—a violation of the implicit social contract between businesses and workers. And they will ask for something like the Posted Workers Directive that puts limits on how long telemigrants can be used by companies and how much they have to be paid.

There is nothing new about these examples of shelterism. Shelterism has a long history of protecting politically powerful industries.

Historical Shelterism: Red Flag Laws and Featherbedding

In the nineteenth century, it wasn't self-driving cars threatening to throw millions out of work, it was human-driven cars. Motor vehicles threatened the livelihoods of many workers in the British horse-drawn carriage and railroad sectors. In Britain, these sectors fought back with reactionary regulations known as the "Red Flag" laws. These laws, which spread to some US states, were as ludicrous as they were effective in slowing automation.

---

19. Jean-Claude Juncker, "A New Start for Europe," Opening Statement in the European Parliament Plenary Session, July 15, 2014.

The most famous, the Locomotive Act of 1865, imposed extreme conditions on "every locomotive propelled by steam or any other than animal power." It required that "at least Three Persons shall be employed to drive or conduct such Locomotive." The "red flag" name comes from the second requirement: "one of such Persons . . . shall precede such Locomotive on Foot by not less than Sixty Yards, and shall carry a Red Flag constantly displayed, and shall warn the Riders and Drivers of Horses of the Approach of such Locomotives."

But the thing that really rendered the new technology uncompetitive was the speed limit: "It shall not be lawful to drive any such Locomotive along any Turnpike Road or public Highway at a greater Speed than Four Miles an Hour," which is walking speed, "or through any City, Town, or Village at a greater Speed than Two Miles an Hour." The laws stifled the automobile industry in Britain for three decades (it was repealed in 1896).

In one of those "truth is stranger than fiction" moments, San Francisco banned self-driving delivery bots from most sidewalks in 2018. The bots allowed are restricted to moving at less than three miles per hour and a human operator must be within 30 feet during testing. In what is either a subtle tribute to historical shelterism, or just a bald coincidence, sidewalk-based delivery bots in Washington D.C. are fitted with red flags to alert pedestrians and drivers.

Automation in cargo handling produced a different type of reactionary regulation—as did the switch from coal-powered to diesel-power train engines. It was called "featherbedding" and forced companies to keep paying workers whose jobs had been rendered obsolete by automation. This seems destined to be copied in future shelterism.

Containerized shipping was a boon to trade and manufacturing from the 1960s. Shipping costs were slashed by the switch to standardized shipping containers that could be loaded and unloaded directly from trains or trucks with massive cranes. The labour- and time-saving technology, however, scuppered the fortunes of highly paid dock workers known as longshoremen, who loaded and unloaded ships using traditional methods.

Ultimately, it was a question of who would bear the economic and social costs of technological change: the workers or the companies. In the

US, longshoremen were unionized and, since they controlled a vital economic chokepoint, they had a good deal of bargaining power. And they used it to get some shelter from the technology. After a series of costly strikes and port blockades, the shipping companies and ports settled the matter by keeping displaced workers on the payroll even when they had little to do. This was called featherbedding.

The situation in the railroad sector was similar and lasted until the 1970s. When the technology switched from coal to diesel, unions managed to force railroads to continue hiring "firemen" even though there was no fire on a diesel train. The laws and contracts that the workers bargained for had names like "full-crew laws" that required a minimum number of workers per train; "train consist laws" that limited the size of trains, and "job protection laws" that required compensation for employees who were laid off or transferred to other duties.

More recently, privacy laws have shielded Swiss financial-sector jobs from offshoring. Switzerland has strict privacy laws for its banking sector. Intentionally revealing client secrets can lead to three years in prison. Unintentional breaches can lead to $250,000 fines. This naturally puts a damper on Swiss banks' enthusiasm for the sort of back-office offshoring that is common in US and UK banks. While the regulation was not designed to protect back-office jobs, it inevitably had that effect. It unintentionally slowed the globotics transformation and sheltered some Swiss workers from globots.

It is easy to think that data privacy laws could be used similarly to hinder the use of telemigrants in many service sectors. Medical, accounting, and data storage sectors could be subject to new regulations justified on the grounds of privacy but politically motivated in a large part by shelterism.

While these sorts of highly specific reactions are inevitable, they will not substantially slow the general rate of job displacement. A very different set of policies could do just that. Many high-income countries have extensive rules, regulations, and laws that govern how and why a worker can be let go—they are called Employment Protection Legislation.

Regulation to Greatly Slow the Globotic Transformation

Micaela Pallini runs a 137-year-old company that thrives on one of Italy's greatest strengthens—its food culture. In the summer of 2012, she passed on a chance to double production via a joint venture. "We didn't pursue it. If the venture failed, Italian laws make it almost impossible to cut our work force to adjust costs."[20]

Italian levels of workplace shelterism are unheard of in the US, but quite common in Europe and other high-income economies. The policy goal of these laws is to protect workers. Or more precisely, to ensure that workers are not the only ones to bear the cost of changes. In some countries, like Britain, the laws are viewed as a matter of basic justice. Workers should not be dismissed arbitrarily and generally speaking they should get some compensation when they are dismissed.

In southern Europe, the laws are aimed at creating a system of life-time employment in that they make it very expensive, slow, and difficult to fire a worker for any reason. Court cases often take years to resolve. The Pallini example illustrates why most economists oppose such sweeping restrictions. As Pallini pointed out, big restraints on firing mean big restraints on hiring. When growth was booming in the pre-1973 decades, these sorts of laws didn't really do much harm. Most firms were growing and hiring since sales were growing. But now that growth rates are much lower, strict Employment Protection Legislation is having pernicious effects on productivity growth. The laws make it very hard for companies to adjust to changing technologies, demand patterns and the like, but such adjustment is the only way to keep productivity growing. Growth requires change and change causes pain. Countries need to find ways to share the pain, but trying to stop the pain by stopping the change will lead to stagnation.

But what if slowing down progress became critical to avoid violent backlashes and social turmoil?

---

20. Quotes from Liz Alderman, "Italy Wrestles With Rewriting Its Stifling Labor Laws", *New York Times*, August 10, 2012.

The most obvious way to slow the advance of globots would be to make it harder, slower, or more expensive for companies to get rid of the workers. In principle, it could be linked to globot-induced firings, but in practice operationalizing this sort of conditionality would be very difficult and time-consuming.

Adding such frictions to the economy is likely to be costly in terms of productivity growth, and it would certainly slow down the march of measured labour productivity. It is thus not a set of policies to implement lightly. Yet, if politicians decide they need to slow down the speed of job displacement, Employment Protection Legislation is one way they could do it. Indeed, since most advanced economies outside the US already have extensive regulatory institutions in place to deal with worker dismissals, this policy option could be dialed up rather quickly.

FROM BACKLASH TO RESOLUTION

Making dramatic predictions about the future is an old business—dating at least as far back as ancient Greek times when the famous shepherd from Aesop's tale cried wolf. But it is worth remembering how that tale ends. There were a few false alarms, but the wolf did eventually come. As the villagers were comfortably ignoring the shepherd's shouts, the wolf destroyed the whole flock. That was their "holy cow" moment (although perhaps they thought of it as their "holy sheep" moment).

At the time this book went to press, there were no signs that digitech would lead to violent reactions. Straightlining the future suggests that it should stay that way and that the changes will come slowly. It is also pos- sible that widespread shelterism and reactionary regulation could slow the impact of digitech's job displacement in ways that allow job creation to keep up. But it is important to keep in mind that things could get out of hand if globots cast hundreds of millions of lives into disarray.

With or without dramatic predictions, the future will arrive. The skills of AI-trained robots and the talents of foreign freelancers will—when combined with their very low costs—take over many of the tasks that humans currently do. Reactionary regulation, or a more violent uprising, may slow the trend, but it is unlikely to postpone it indefinitely. There will be a resolution. If we do make it to the long run, we are likely to find ourselves in a much better society.

9

Globotics Resolution: A More Human, More Local Future

Amelia, the white-collar robot we met in Chapter 1, is making jobs as well as taking jobs. One crazy job that Amelia created was for Lauren Hayes—the real woman on which Amelia's avatar is based. Since Amelia is known to millions, Hayes—a twenty-something model—is a celebrity in a strange way. An executive from a large insurance company that uses Amelia told Hayes that his sixty-five thousand employees loved her. Hayes herself, by contrast, was not a natural fan of Amelia from the start.

"It was really creepy," she said. "I didn't imagine it would be so realistic. I didn't realize it would talk or have motion." When the human model had her first photo session for the digital model, Hayes worked out that being the human face for a white-collar robot would be a very odd job. As Hayes put it, "At that moment, I was like, this is not like anything I've ever done before. This is not a print job for Gap."[1] To capture 3D images and natural body positions and facial expressions, the photo shoot used something that looked like the *Star Wars* Death Star but turned inside out.

There are lessons to be learned from this crazy job. Hayes's job depended upon her humanity. As a matter of pure logic, many of our jobs in the future will look more like Hayes's than we think.

---

1. Quotes from Sarah Kessler, "Inside the Bizarre Human Job of Being the Face for Artificial Intelligence," *Quartz.com*, June 5, 2017.

Few Americans and Europeans will be able to compete with globots. This in turn means we won't. Globots will do what they can do. We will do the work that globots can't do.

There is no use in thinking about which jobs these will be. If history is a guide, they will mostly be in sectors that we haven't imagined, as labor economist David Autor points out.[2] But although we can't know what the jobs will be called, we can build intuition for what they will be like. We can do this by studying the things that humans do better than robots and telemigrants. The place to start is a deeper look at humanity's unique talents.

WHEN IS HUMANITY AN EDGE OVER SOFTWARE ROBOTS?

Humans have unique advantages over AI-trained computers in things like judgment, empathy, intuition, and comprehension of complex interactions among teams of humans. Psychologist call this "social cognition," and we have it for very specific, very deep-seated reasons. It provided an evolutionary advantage.

Compared to other large animals, Homo sapiens are particularly underwhelming in the tooth, claw, and muscle departments. Nevertheless, we are the ones that bestride the planet—having wiped out, tamed, or enclosed a slew of species that could—in a one-on-one fight—beat the living daylights out of us. This roaring success as a species is due to our social brilliance.

The reasons that humans study chimps who live in cages rather than the other way around is that people can band together and do amazing things. Social cognition is the key that opens the door to this very human skill. Social cognition means being able to conceptualize what is going on inside the minds of others, to understand what's going on inside your own

2. David Autor, "Why Are There Still So Many Jobs? The History and Future of Workplace Automation," *Journal of Economic Perspectives* 29, no. 3 (Summer 2015): 3–30.

mind, and to loop back and comprehend how others are thinking about what you are thinking. This was critical to humans' survival.

As Michael Tomasello wrote in his pathbreaking book, *The Cultural Origin of Human Cognition*, social cognition allowed humans to live in relatively large groups where survival turned on the ability of individuals to cooperate with and manipulate others within a complex web of relationships involving trust, kinship, and dominance. The equipment for this is hardwired into everyone's brain. One element of the wiring is called "social mirroring."

When we interact with others, we communicate intentions and feelings along with more businesslike information. We put the facts into context using gestures, facial expressions, body postures, and the like. One part of our brain—the "mirror neurons"—are devoted to this social interaction. Rather than "monkey see, monkey do" mechanisms, these are "people see, people feel" mechanisms.

Marco Iacoboni, who has the that's-a-mouthful title of professor of psychiatry and biobehavioral sciences, explains it this way: "When I see you smiling, my mirror neurons for smiling fire up, too, initiating a cascade of neural activity that evokes the feeling we typically associate with a smile." What this means, he adds, is that "I don't need to make any inference on what you are feeling, I experience immediately and effortlessly (in a milder form, of course) what you are experiencing."[3] All this is instantaneous and effortless, and we are rarely aware of it, although you can often see it in how people talking together unconsciously synchronize their head nods, arm-crossing, hand gestures, and the like. The most sensitive among us can feel physically ill when they see others experience violence or disturbing emotions. Hearing a sad story makes us feel sad, maybe even cry, even if it happened long ago to someone far away. Mirror neurons turn sound waves into emotions.

In short, a big part of the human brain is hardwired for social intelligence. Not all of us are equally good at social cognition, just as we aren't all

---

3. Quotes from Jonah Lehrer, "The Mirror Neuron Revolution: Explaining What Makes Humans Social," *Mind Matters* (blog), ScientificAmerican.com, July 1, 2008.

equally good at algebra. But as it turns out, computers are much better at algebra than they are at social cognition, and this provides an edge that will allow humans to stay competitive in jobs that involved social interaction.

Why AI-Trained Computers Have Trouble with Social Cognition

Some AI-trained computers can quite accurately judge the emotions of humans they are interacting with on a one-on-one basis. We met one, Ellie the AI-trained robo-therapist, in Chapter 6. There are even robots that have learned how to elicit emotions from humans, like trust and sympathy. A therapeutic robot named Paro, for example, looks like a baby seal. It has been providing company and comfort for elderly Japanese since 2012. But this is a long way from understanding group dynamics.

Understanding what is going on in a group requires us to understand how each team member is feeling. Psychologists have a rather strange name for this: "theory of the mind." By this they mean the capacity to identify feelings, beliefs, intents, desires, and falsehoods in others since we have a model of other people's minds in our own mind. Just think about how you know how your mother, spouse, or child will think about something you are thinking about doing. You "know" how they'll react because you have a model of how they think tucked somewhere between your ears. There are many loops and levels in this process—something like the 2010 sci-fi Hollywood thriller, *Inception*.

The first level is to understand what others are thinking or feeling. The second level is to understand how we feel about each team member and how they are feeling about us. If the group is to get along, we usually have to understand how each team member feels about the other members. That's the third level. Really successful managers and team members often go a few more levels up in terms of understanding what others are understanding about each other's understanding. Computationally, this is a problem that gets extremely difficult as the numbers rise. The branch of mathematics that studies this sort of thing is called combinatorics.

The key point is that the number of possible combinations grows extremely fast with the number of things that can be combined. Consider a case of three people. Using just first-level social cognition, Ms. 1 needs to understand two things—what Mr. 2 and Mr. 3 think. But suppose what Mr. 2 thinks depends on what he thinks Ms. 1 and Mr. 3 are thinking? Then the leader has to also understand Mr. 2's view of Ms. 1's and Mr. 3's thinking, and likely enough Mr. 3's thinking about Ms. 1 and Mr. 2's thinking. As higher levels of social cognition are required, the amount of social thinking goes through the roof, especially as the number of team members rises, and the range of possible views expands.

Despite the complexity, many of us can do this social math instantly and without conscious thought. Normal children, for instance, reach the first level by four years old and the second level by six years old.

This type of social brilliance is one of the evolutionary gifts bestowed on us by hundreds of thousands of years of evolutionary selection in a world where humans were viewed as food by more physically capable species. Our edge over white-collar robots is our innate embrace of team-building practices like fairness and reciprocity, and empathy and impulse control. Most of us actually enjoy working cooperatively. In short, humans are social-math geniuses; computers aren't.

A second critical workplace skill that arose from evolutionary pressure is the ability to detect cheating and assign trust.

Social cooperation slips very quickly into social exploitation and free riding. If you worry about yourself when everyone else is worrying about the collective good, you are likely to thrive if the others cannot detect your cheating. But, in fact, many of us have incredible mental powers in the cheating-detection department. We have very finely honed but unconscious ways of telling if someone is lying. Part and parcel with this is a deep-seated abhorrence of exploitation, on one hand, and a deep-seated abhorrence of social exclusion on the other. The pair generates social behavior that fosters cooperation and trust.[4]

---

4. For a textbook exposition of these social psychology concepts, see Graham M. Vaughan and Michael A. Hogg, *Social Psychology*, 7th ed. (London: Pearson, 2013).

Machine learning has problems with this social cognition for a few reasons. The first is that even today, computers aren't powerful enough. The second is that we don't have the right kind of data. The third is more speculative. Machine-learning techniques are a shallow imitation of the biology of human thinking and learning, so it may be that a whole new computer-science approach is needed if machines are to rival humans in the most human skills.[5]

Algorithms Are Too Small and Too Blunt for Social Cognition

The mainline AI technology that is driving service-sector automation is machine learning, as we saw. One of the main approaches used is called an "artificial neural network." This consists of artificial neurons, connections among them, and the weights given to the various connections. Each neuron can be thought of as a tiny computer that tackles a tiny part of the problem under study—say, recognizing a song or face. The connections and weights are essential since they coordinate the overall problem solving. These work roughly like the human brain, but only very roughly. And they are much, much smaller.

In 2017, a typical neural network had at most millions of artificial neurons.[6] A typical human brain has a thousand times more neurons and several hundred trillion connections between them. Moreover, artificial neural networks have fixed connections among the neurons. In the human brain, the connections adapt to cognitive needs. In artificial networks, the messages are in digital form—it is "on" or "off," "yes" or "no." Biological neural networks are subtler.

As the neuroscientist Christopher Chatham puts it: "Accurate biological models of the brain would have to include some 225 million billion

5. See Brenden M. Lake, Tomer D. Ullman, Joshua B. Tenenbaum, and Samuel J. Gershman, "Building Machines That Learn and Think Like People," *Behavioral and Brain Sciences* 40 (2017).

6. Sean Noah, "Machine Yearning: The Rise of Thoughtful Machines," *Knowing Neurons* (blog), KnowingNeurons.com, April 11, 2018.

interactions between cell types," and a list of highly technical sounding brain-bits. "Because the brain is nonlinear, and because it is so much larger than all current computers, it seems likely that it functions in a completely different fashion."[7]

In the human brain, messages are transmitted when nerve cells (neurons) "fire". To fire in this sense means to pass a weak electrical pulse from one end of the nerve cell to the other. But in the brain, the message transmitted depends on the speed of "firing" and the synchronicity with which groups of neurons fire. Moreover, the human brain is massively parallel in the sense that it is solving many problems at the same time (mostly unconsciously). Artificial neural networks are, by contrast, modular and serial. For example, a photo is fed into the computer and it determines whether there is a face in the picture.

The point of this is that social intelligence is something that will prove valuable in the competition with white-collar robots. Many of us are socially brilliant. Better yet from a social stability perspective, it is not necessarily the most educated among us that have these talents.

What Else Can't Machine Learning Learn?

AI is really just data-based pattern recognition, and pattern recognition is not intelligence. AI is thus not intelligence in the broad sense of the word that psychologists use. White-collar robots trained by machine learning do not have a capacity to think; they cannot reason, plan, or solve problems they have not seen before; and they cannot think abstractly or comprehend complex ideas that are more than patterns in data.

Computer scientists may eventually find ways to give white-collar robots general intelligence, but that is a long way off—and most definitely not a clear and pressing problem for Europe's and America's middle class.

---

7. Chris Chatham, "10 Important Differences Between Brains and Computers," *ScienceBlogs*, ScienceBlogs.com, March 27, 2007. For a more recent discussion, see Lance Whitney, "Are Computers Already Smarter Than Humans?" *Time Magazine*, September 29, 2017.

One key limitation is data. AI pattern recognition is usually based on structured data—data where the questions and answers are clear. In many social situations, neither the questions nor the answers are clear. That's why they are called feelings rather than thoughts. This matters since AI computers are uncannily good at recognizing patterns—but only specific ones. This is why when Amelia and her kind can't find a match, they kick the case over to someone who has real intelligence.

Humans are, and are likely to remain, better than white-collar robots in activities that involve situations where the issues are unclear, success is hard to define, or the outcomes are unclear. Likewise, AI can't learn without masses of data, so chores where there is little data are also likely to remain in human hands. By contrast, AI will very soon be a serious competitor for the aspects of our jobs that can be codified with a massive data set.

There are two deeper limitations to the computerization of human activities. The first is called the "black box" problem, or the issue of responsibility for decisions taken.

Personal Responsibility—Black Box Problems

One futurist, the billionaire Vinod Khosla, boldly predicted that "computers will replace 80 percent of what doctors do" because computers would be cheaper, more accurate, and more objective than the average doctor. That was back in 2012 and the prediction is not looking good.

Based on analysis and recent trends, the US's job counter, the Bureau of Labor Statistics, projects the number of doctor-jobs will grow by 14 percent per year up to 2024. More recently Khosla said: "I can't imagine why a human oncologist would add value, given the amount of data in oncology." He thinks AI will have eliminated human radiologists in five years.[8] Well, maybe, but there are some problems.

8. Liat Clark, "Vinod Khosla: Machines Will Replace 80 Percent of Doctors," *wired.com*, September 4, 2012.

When an white-collar robot does a clever thing, like recognize a face, it is using a very large statistical model to find patterns in the data you gave it—in this instance, a photo. AI programs don't calculate; they make educated guesses. AI is not looking for an exact answer like it would if you asked it to calculate the number of days you've been alive. AI programs guess. These algorithms are not unlike the models that weather forecasters use every day. Weather forecasters plug in a huge number of weather factors, and the computer model spits out a guess about what the weather is likely to be. This guessing feature is why Facebook sometimes tags the wrong people in your photos. It is also probably why AI programs—like Poppy and Henry who we met in Chapter 4—seem a lot more "human" than Excel spreadsheets, even though all of them are just software.

One big limitation—called the black box problem—is that the algorithms that generate the guesses cannot explain why they guessed what they did. The statistical models are not set up to explain themselves. When IBM's Watson made the life-saving call for the Japanese leukemia patient, for example, the doctors could not know what exactly tipped off the computer model. Likewise, when Google Translate does its "thing," it cannot explain why it used one word instead of another.

This matters in many settings. It means that many jobs cannot be completely replaced by a computer algorithm making guesses—even one that guesses better than the average human. For instance, would you allow a super-accurate computer doctor to decide to amputate your right leg when the computer could not answer your why-is-it-necessary questions? This feature means AI systems will, in many instances, work alongside humans who can take responsibility for their decisions.

In the end, this means that it will be very hard to computerize jobs where someone has to be held accountable for the decisions made, or where the humans using the guesses want to hear the reasoning behind it. This is probably a point that will keep many high-level professionals in business, even if there will be fewer of them. When it comes to decisions ranging from architectural design to medicine and the selection of art, people will want to know "why," not just "what." And they'll want to hold someone accountable if the wrong decision is made.

A second, deep limitation of the machine learning approach is something that is well known to economists in a different setting.

The Lucas Critique of AI-Trained Algorithms

Nobel Prize winner economist Bob Lucas famously explained why Keynesian economic models, which used to work well in the 1960s, fell apart in the 1970s when inflation picked up. His point—the so-called Lucas critique—was that the models weren't describing how the economy actually worked. They were describing how it worked as long as some unexplained correlations continued to hold. The exact details aren't important here, but the basic point is.

Algorithms only work as long as the correlations that existed in the training data continue to hold. If something fundamental shifts and this leads the correlations to break down, the guesses based on the correlations could go haywire.

To take a simple hypothetical example, suppose you trained a software robot to distinguish boys from girls in 1950s school photos. One of the factors the algorithm would almost surely pick up on is hair length, since almost every girl had longer hair than almost every boy back then. Note that this importance of hairdos would not be explicit—you probably couldn't even be sure if it was baked into the algorithm. In the 1960s and 1970s, something fundamental changed that led many boys to have longer hair and many girls to have shorter hair. Using the 1950s algorithm would thus misclassify many students.

The topline here is that AI-trained robots do not understand the world. They just understand patterns in their training data sets. This reliance on correlation rather than causation will inevitably lead to very systematic mistakes when underlying factors change.

This is another reason AI robots are unlikely to be trusted with critical tasks. There is no danger in letting them suggests tags for your Facebook friends. There could be real danger if we fully relied on them for more essential tasks. There will long be a demand for having humans in the decision loop.

So what does this mean for the future of work? What type of work will be naturally sheltered from AI competition? These are very difficult questions to address given the radical diversity of occupations. To make headway we have to simplify to clarify.

WHICH ACTIVITIES WILL BE SHELTERED FROM AI-LED AUTOMATION?

Every occupation involves a whole pile of tasks. Some of these tasks are things that robots are good at and some are tasks at which robots are useless. The Oxford scholars behind the most influential study of AI automation—Carl Frey and Michael Osborne—argue that the hardest tasks for white-collar robots involve creative intelligence and, as discussed, social intelligence.

Creative intelligence means being able to devise new, good ideas and solutions. By social intelligence, Frey and Osborne mean being aware of people's reactions to events and being able to react appropriately. Typical workplace tasks that draw on social intelligence are negotiation (getting people to cooperate and reconcile differences) and persuasion (getting people to agree on ideas, ways of doing things, etc). It is also important in tasks like assisting and caring for people, providing emotional support, and the like. The parts of jobs which rely heavily on creative and social intelligence are likely to remain sheltered from robots in coming years.

A related approach to the "which jobs will be sheltered from robots" question was taken in 2017 by the experts at McKinsey consulting firm in an important study, *A Future that Works: Automation, Employment, and Productivity*. The approach focused on what we do in our jobs rather than on what is done in any specific job.

This approach involved a few steps. First, they classified, into eighteen workplace "capabilities," all the things that workers need to do in all jobs (these are the capabilities we saw in Chapter 6). Then experts judged how good today's AI is at each of these eighteen. To bring this from "capabilities" to jobs, they classified all workplace chores into seven "building-block" activities. These are: doing predictable physical activities, processing data, collecting data, doing unpredictable physical activities, interfacing with stakeholders, applying expertise, and managing and developing people. To judge the importance of each of these seven activities, they calculated how much time is spent on each of the seven activities looking across all US jobs. In Figure 9.1, the results are shown with the light bars. For example, 18 percent of time at work—adding up across all US workers and all US jobs—is spent on predictable physical activities.

The last step was to cross-match the eighteen capabilities and how automatable they are with the seven activities. The results, illustrated with the black bars in the Figure 9.1, show the share of each building-block activity that can be automated. So what do the McKinsey calculations tell us?

The least automatable activity is "managing and developing people." This is an activity that fills about 7 percent of all the hours worked in the US and 9 percent of it is automatable. This is quite in line with humanity's edge. Managing involves lots of emotional and social skills, as well as

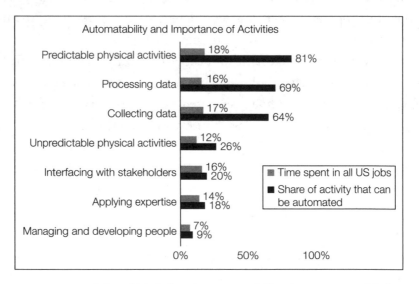

Figure 9.1 Automatability of Workplace Activities and Their Importance in Work. SOURCE: Author's elaboration of data published by McKinsey Global Institute, Exhibit E3. "A Future That Works: Automation, Employment, and Productivity," January 2017.

dealing with groups of people. Since computers are bad at both of these, jobs involving lots of managing and fostering of people are likely to be shielded from automation.

The next least automatable activity is "applying expertise." Again, this lines up with skills where humans have an edge over software robots—at least in a subtle way. It is true that software robots are already very good at mastering large amounts of data. Think of Amelia's ability to learn a two-hundred-page manual on SEB banking procedures for opening accounts, or legal-bots that can read through and classify mountains of decisions written by judges. But knowing things and applying the knowledge are two very different things.

The AI-trained bots in these cases are really something like a talking encyclopedia—you can ask them questions and get great, clear, history-based answers, but they don't—by themselves—know what questions to ask. The point is that applying knowledge involves recognizing ill-defined patterns and issues in new cases. Jobs that involve applying experience-based expertise will be sheltered. The jobs under threat are those held by humans who are today assisting these experts. Another aspect of AI that strengthens this conclusion is the black box and personal responsibility problems. AI cannot take responsibility, but in many cases the people asking for the advice want to be sure that they can hold someone responsible if the advice doesn't work out. And it's not just the clients. The law will want to be sure there is accountability.

The next activity, "interfacing with stakeholders," is only about 20 percent automatable. This sort of activity plays to the social brilliance of humans and against the cognitive strengths of AI-trained white-collar robots. These "soft," human-side jobs are surely some of those that will be sheltered from the rapid job displacement, although it is likely that some local humans will be replaced by online humans telecommuting from afar.

The fourth difficult-to-automate activity is unpredictable physical tasks—this covers things ranging from dentistry to bonsai gardening. While some of these may eventually be done by robots controlled by remote humans (called telerobots), it seems that many of these jobs will be sheltered in the coming years.

The other three activity groups (predictable physical activities, processing data, and collecting data) are far more automatable. Jobs that involve a lot of these activities will see a lot of job displacement in the near future. The most "at-risk" activity is performing physical activity and operating machinery in predictable environments. Over 80 percent of the hours spent on such activities draw on skills that can be automated by AI-trained robots. While not all such activities in all jobs will be replaced, this is the sort of activity that will experience disruption.

Here "automatable" means the activity could, technically speaking, be automated. How fast they are automated in practice is a question that is much harder to answer. The reason is that the answer turns on business decisions and these, in turn, depend on how each firm thinks about what their competitors will do. It is exactly this sort of herd behavior that makes the timing difficult to predict. But it also means that when the automation does start, a cost-cutting race could wildly accelerate the process.

This discussion of automatable activities is insightful, but not fully satisfying. It is great to know that many of us will be working in jobs that don't yet exist and to know what sorts of things we'll be doing in these jobs. But we all care about the jobs we have today. We all want to know whether our own occupation is likely to be affected. This is why it is instructive to map the capabilities of AI into real occupations. McKinsey has done this for us.

Which Occupations Are the Most Sheltered?

McKinsey classified all jobs into nineteen different sectors. They then used their underlying estimates of the automatability of capabilities to generate an estimate of the share of hours that are automatable in each of the sectors. Focusing only on the sixteen services they list, the findings are plotted in Figure 9.2.[9]

9. The three omitted catagories of jobs, and share of work that is automatable (in parentheses) are: manufacturing (60 percent), mining (51 percent), and agriculture (57 percent).

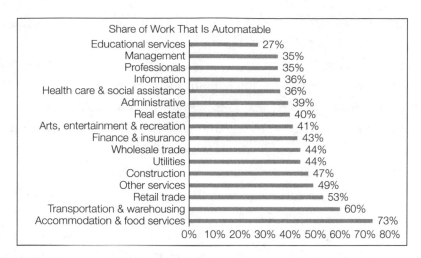

Figure 9.2 Share of Work That Is Automatable in Service Occupations.
SOURCE: Author's elaboration of data published by McKinsey Global Institute, "A Future
That Works: Automation, Employment, and Productivity," January 2017.

How should we think about these estimates? Plainly, there are many jobs
in, for example, the "accommodation and food services" sector that will
not be automated since they involve the sorts of activities that computers
are not good at. But in a rough way, it suggests that a substantial fraction—
up to 73 percent—of the hours now put in by humans in this sector will, in
coming years, be replaced by robots. That is a lot of jobs.

On the sheltered side, less than half the tasks are automatable in jobs
like education, the professions (lawyers, accountants, architects, etc.),
management, and healthcare and social services. These tend to be jobs
that involve lots of judgment, emotional intelligence, and dealing with unexpected situations.

The Oxford professors, Frey and Osbourne, take a somewhat different approach but come to pretty similar conclusions. The most sheltered occupations include accommodation service managers, elementary
school and kindergarten teachers, dietitians and nutritionists, occupational therapists, dentists, general practitioners and family physicians,
specialist physicians, fire chiefs and senior firefighting officers, denturists,
audiologists and speech-language pathologists, textile patternmakers,

leather and fur product makers, and outdoor sport and recreational guides.

This list is both fascinating and useless for most people. After all, how many fire chiefs can there actually be in the world? But the point of the list is not to highlight particular jobs but rather to give a flavor of the sorts of jobs—many of them completely unimaginable today—that will employ most people. More generally, the sectors in which at least 40 percent of the occupations are shielded from AI included: management; education; professional, scientific, and technical; media, arts, entertainment, and recreation; government; and utilities.

The broad answer to the "which-jobs-will-be-sheltered question" is rather clear when we combine the McKinsey and Frey-Osbourne estimates. The protected jobs will be those that stress more human features: caring, sharing, understanding, creating, empathizing, innovating, and managing.

How long will this natural "human-edge" provide shelter from the globots? The points made previously about the general limits of machine learning suggest that the shelter will last a long time. The McKinsey experts have provided more precise estimates.

When Will Computers Learn the Most Human Skills?

Machines have not been very successful at acquiring social skills. But AI is advancing rapidly. If jobs and activities are to remain sheltered from automation, we need to look at projections of how soon machines will attain human-level performance in the skills they are not yet good at. Again the McKinsey experts have done the heavy lifting in a unique way.

The McKinsey A-team on this includes economists, business strategists, and AI scientists. Drawing on a broad range of expertise, including those who helped quantify the current capacity of AI, they peered into their crystal ball to see when white-collar robots will acquire human-like skills. What they find is encouraging from the perspective of social stability.

For the most human-like tasks—especially those involving social cognition—they suggest that AI skills will remain below that of an average human for the foreseeable future. They estimate that it will take something like fifty years for AI to attain top-level human performance in the four social skills that are useful in the workplace: social and emotional reasoning, coordination with many people, acting in emotionally appropriate ways, and social and emotional sensing.

Making projections that far out takes a brave soul, but it is necessary that someone be brave. Society has to make choices about things that will have effects that last decades, like educational and regulatory systems. Businesses have to make long-term choices about staffing and strategies. The bottom line is that social skills are likely to remain sheltered from AI competition for much of our lifetimes. Much the same can be said for other skills that are not easily codified. Figure 9.3 puts numbers to these guesses.

Figure 9.3 shows the estimated year by which AI will attain top-level human performance in the eighteen different workplace skills listed. What is remarkable is that in three of the six thinking skills, AI is already more capable than the average human, but in the others, humans are projected to have an edge for a very long time. In "logical reasoning and solving unknown problems," humans should have the upper hand for another forty years. For creativity, it is fifty years, and for "generating novel patterns, or classifying new situations into new categories" it is twenty-five years.

Many of the limitations of AI have to do with the social hardwiring of the human brain. Telemigrants, being human, do not suffer from these shortcomings. Telemigrants possess the same sort of social, emotional, and creative intelligence as local humans. Yet telemigrants have their own limitations. There are certain workplace tasks that require real people to be in the same room at the same time. This reality leads us to a different set of considerations.

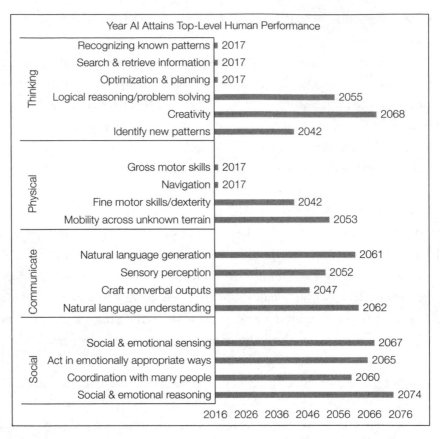

Figure 9.3 The Year AI Attains Top-Level Human Performance in Workplace Skills. SOURCE: Author's elaboration of data published by McKinsey Global Institute, "A Future That Works: Automation, Employment, and Productivity," January 2017.

WHEN IS BEING LOCAL AN EDGE?

In looking for the sort of jobs that will be naturally sheltered from telemigrants, we need to think hard about why face-to-face communication matters. Since verbal communication is almost costless these days, a good place to start is nonverbal communication. This is a fascinating area that is widely studied by psychologists.

Nonverbal Communication

Communication is more than just words. When people are face to face in the same room, psychological experiments suggest that less than 30 percent of the information exchanged stems from the words spoken—some communication researchers put the number as low as 7 percent. The rest is nonverbal. Reflect upon the oddness of this factoid.

Why should it matter so much that you are looking at someone while you're hearing them? The answer is as simple as it is profound. It has to do with the key role that communication played in human evolution and the fact humans and our ancestral species communicated nonverbally—much as apes do today. It is fascinating stuff.

Nonverbal communication is far more ancient than spoken communication. It is deeply baked into our brain circuitry by evolution for the simple reason that humanoids have not been speaking for that long. Humanoids started speaking somewhere between fifty thousand and two hundred thousand years ago (some say earlier), yet humans split off from the other great apes about six million years ago (give or take a million years—this isn't rocket science).

For millions of years being talkative didn't involve talking. Nonverbal communication was the best we could do. Humanoids "talked" with facial expressions and other forms of body language. This is still how it is for most nonhuman apes. Indeed, if you have ever watched monkeys at the zoo, you'll realize you can actually understand some of what they are "saying" to each other. They share some of our facial expressions (or is it the other way around?).

The key point here is that the ability to send and receive nonverbal messages was an important element of the "survival of the fittest" long before spoken language. That is why our brains are hardwired for nonverbal communication. This has an important implication for telemigration.

The nonverbal signals we send are more authentic, and thus more trustworthy, exactly because they are more innate, and far more deeply embedded in our brains than are words. For example, while languages differ a lot across the world, nonverbal communication is pretty universal.

Experiments from around the world have identified six basic expressions that are universally understood: disgust, fear, joy, surprise, sadness, and anger. Some of them are so innate that children who are born blind use them. And you surely use them unconsciously even when speaking to someone on the phone.

One key point is that nonverbal communication provides a very rich "dictionary" of expressions. This unspoken messaging involves far more than the face, but the face is the focal point of it. There are over forty muscles in the human face (a surprisingly large share of the six hundred or so in the whole body). With 40 muscles to play with, the number of possible combinations is almost countless.

Another thing to keep in mind is that much of the information-processing related to reading these expressions happens without us knowing about it. Unlike our conscious brain (what we called System 2 before), the unconscious brain (System 1) is very, very good at multitasking. It can process large amounts of visual and audio data almost instantly, and effortlessly. This sort of thinking is what generates "gut reactions" and "intuition" about people's true intent or trustworthiness. Our brain figures this out without asking our permission.

The lack of conscious thinking is one of the reasons that you probably have not given much thought to why, for example, Facetime or other video calls with loved ones are so much more satisfying than regular phone calls. Or why it is easier to say no to someone by email than it is to do in person.

One set of nonverbal messages that do not come through on standard video calls are known as "microexpressions." They get their name from the fact that they last only 1/25 of a second. These split-second facial changes provide important clues as to whether a person is concealing an emotion, consciously or unconsciously.

Microexpressions are one of the reasons face-to-face meetings generally lead to better understanding and trust than phone calls or Skype. Regular video-conferencing equipment doesn't have resolution that is good enough for people to see microexpressions. If you've ever watched a movie in both its SD and 4K versions, you'll see how much more "talkative" facial expressions are when the resolution is almost lifelike as it is with 4K.

Humans' Hardwired Social Decoder

The flip side of this unconscious communication is equally important for understanding why face to face matters. Decoding nonverbal messages is hardwired into our cerebral circuitry, but so too is the sending of nonverbal messages. These go out unconsciously, rapidly, and in ways that are hard to control.

If you've ever tried acting, and you're not Meryl Streep or Benedict Cumberbatch, you'll have realized how hard it is to pretend you are feeling emotions that you are not actually feeling. The same but opposite thing happens if you try to pretend that shocking news does not bother you. It is easy to lie with words; it is hard to lie face to face. And it is exactly this unconscious aspect of the messaging that leads us to give it such credence. It is why we tend to trust people more when they say it to our face.

Researchers who focus on this have identified five kinds of nonverbal communication: body language (kinesics), touching (haptics), voice quality (vocalics), physical proximity and relative positioning of speakers and listeners (proxemics), and timing (chronemics, for example, how long different speakers speak).

Body language is one of the best known of these. It is a key reason that talking in person is a much more effective way to establish trust and ensure cooperation. Body language covers things like gestures, head movements, posture, eye contact, and facial expressions. These movements are widely appreciated as sending important signals. But there are some subtleties that help us think about why real face-to-face exchanges are more effective.

A key judgment we all make when dealing with people is whether we can trust them. The ways we do this are very reliant on nonverbal clues. People can "read your face" for clues as to whether you are trying to deceive or mislead them, or whether your words really reflect your intentions. But it is not just the face. When young children lie, for example, it is their body movements that usually give them away.

Psychologists who study lying call this mismatch "leakage." That is, when people are trying to mislead, they have trouble getting all their verbal and nonverbal signals to "say" the same thing. Often their true intent "leaks" out via the kinesics—in facial expressions (breaking eye contact), gestures (touching the face, crossing the arms, swinging legs), or the tone of voice.

But there is nothing routine or automatic about this. There is no "I am lying" muscle that twitches every time you tell a ripe one. Instead, experts look for incongruous clusters of expressions and microexpressions that indicate some leakage is going on. This suggests that the verbal message does not reflect what is really going on in the speaker's head.

Even very good liars have trouble stopping "microexpressions." The main microexpressions involve rapid and small movement of the lips, eye brows, eye lids, wrinkles around the nose and other facial muscles.

Microexpressions are critical when thinking about how easily telemigrants can fit into the office, so they are worth looking at a bit more. It is worth watching some of the many YouTube videos that analyze microexpressions on the faces of famous people telling lies. Watching a five-minute video will convince you more than reading a whole chapter on it (due to the power of nonverbal communication, of course). My favorite video shows a slow-motion analysis of Lance Armstrong when he denied taking performance-enhancing drugs in a TV interview.

Studies show that it is easier for liars to control their facial expressions than their arms, legs, and posture, so the face is only part of the equation. Researchers have found that facial expressions are the easiest to control and thus the least reliable of the various forms of body language. Your body movements are less controllable, and your voice is the least controllable of all. This is why many speakers find it easier to hide behind a lectern, or desk—they don't have to worry about controlling the messages that are being sent by the lower body language.

Another obvious advantage of local people over remote people is local knowledge. This is not immutable.

Local Knowledge

Andrew Marantz wanted a job with an Indian call center. The first step in the process was a three-week training course aimed at neutralizing his Indian accent and training him to avoid uniquely Indian English words and expressions. The second step was an immersive course in local culture. This step involved things ranging from memorizing idioms and US state capitals to watching *Seinfeld*, and eating burgers and pizza.

Marantz's training for phone conversations illustrates the important, if obvious, fact that it is easier to communicate with and trust people who share your culture. Some of this is pure mechanics. People from the US have a very hard time understanding most people from Glasgow. Some of it has to do with trust. In Switzerland, for example, strangers who speak a Swiss German dialect are much more readily trusted by Swiss Germans since the dialect indicates a childhood spent in a culture where rules are known and respected. This sort of clannishness can also be traced back to its evolutionary roots—which is why it is so prevalent and obvious in today's world.

The importance of local knowledge is not equally important in all tasks. When it comes to getting instructions on how to, say, restore your hard drive from Dropbox, local culture is not first on the excellence list—technical capacity and patience are far more important. But for a psychotherapist, a key part of the job is really understanding the patient, and here it helps to have a very advanced understanding of the environment where he or she was raised.

Which Jobs Will Be Sheltered from Telemigrants?

Given the vast wage advantage that foreign workers have over those sitting in the US, Europe, Japan, and other advanced economies, sheltered jobs will be those that involve things that just cannot be done from far away. Intuitively, these are jobs where it is important to actually be in front of a particular piece of equipment, to be in the room with co-workers or clients, or to be in a particular place.

A decade ago, Princeton professor Alan Blinder classified jobs according to these basic criteria by examining the job descriptions listed by the US government, as noted in Chapter 5. He found that a very large number of occupations in the US had to be performed in a particular place. These, he judged, were immune to competition from remote workers. Examples include farmers who have to be on the farm, child caregivers who have to be with the child, and attendants at Disneyland who just have to be there to get the job done.[10]

While the telecommuting technology has improved enormously since Blinder did his work, and far more people work remotely, this have-to-be-there feature of a job is still an effective shield from foreign online competition. But what about jobs for which you don't absolutely have to be there, but being local provides an advantage? Are these jobs that telemigrants can take?

Blinder teamed up with his Princeton colleague, Alan Krueger, to look at more refined approaches to identifying the jobs that are the most and least exposed to competition from telemigrants. What they did was survey people in the US to find out whether they thought their job could be done remotely. They found that many people did believe their jobs could be offshored.

The sectors where less than 20 percent could be done by telemigrants— were mostly of the have-to-be-there type: jobs in hotels and restaurants, transportation and warehousing, construction, leisure industries, education, and health and social care. The sectors most vulnerable to telemigrants were professional, scientific, and technical sectors; finance and industry; and media sectors. Blinder and Krueger estimated that over half the jobs in these sectors could face direct international wage competition.

Having looked at which types of tasks are likely to be spared from automation by white-collar robots, on the one hand, and having looked at the tasks that will be shielded from remote workers, on the other hand, the next question is obvious. Which task will be shielded from both cognitive computers and foreign freelancers?

10. Alan Blinder, "How Many US Jobs Might Be Offshorable," *World Economics*, 2009.

WHAT JOBS WILL BE SHELTERED FROM AI AND RI?

The index developed by the Oxford professors, the Frey-Osbourne index of automatability, makes it easy to see where AI is not good enough. The index developed by Princeton professor Alan Blinder, the Blinder index of offshorability, does the same for telemigrants. Combining these lets us see which of the current occupations are likely to be immune to both members of the disruptive duo—automation and globalization.

Specifically, I took a list of all the occupations that are listed as not offshorable by Blinder. These are the sorts of work that are not likely to be under threat from remote intelligence (RI). Some of the occupations on this RI-shielded list, however, are highly automatable given current AI capacities. Striking off these AI-exposed occupations from the RI-immune list yields a list that is very interesting for the nature of future of work. The occupations left on the list have a low probability of being displaced by the white-collar robots and a low probability of being replaced by telemigrants. These are today's jobs that are likely to be sheltered in the future.

A couple hundred of the approximately eight hundred occupations count as "sheltered" from AI and RI. Once again, it is useful to point out that most of the jobs of the future will be in occupations that are not on any of today's lists, but the list does highlight the types of jobs that globotics will pass by. More indirectly, this list provides inspiration for thinking about what the new, unknown jobs might look like.

The largest category is made up of management jobs. The list reflects the fact that management usually involves getting people to do things well and fast. Usually that also means getting people to work with each other—all things that involve social intelligence, which AI is bad at, and establishing personal rapport, trust, and motivation, which RI is bad at.

Many occupations related to professional and scientific specializations also come out quite sheltered. These are jobs such as compliance officers, financial examiners, management consultants, event planners, landscape architects, and civil engineers. Again, these are rich in tasks that involve high levels of perception and manipulation, creative intelligence, or social intelligence. Many types of engineers fall into these categories since engineers are typically trying to make things work.

Among the professionals, the key to being sheltered is the requirement of being good at in-person human interaction, or dealing with unstable or unknown situations. These include lawyers, judges, and related workers, and many healthcare professionals. Notably absent from the list are accountants, editors, and lawyers.

Scientists are, almost by definition, dealing with things that are unknown, or very poorly understood, and thus shielded from AI. Many of these scientists have to work in teams, and their work involve the sorts of innovative tasks that are best done when everyone is in the same room.

The social sciences—being people sciences—tend to be sheltered, at least in the case of those that involve interacting with groups of people. The shielded jobs include many types of psychologists. Sociologists, urban and regional planners, anthropologists and archeologists, and political scientists also get high shelter scores. Healthcare service providers are largely sheltered since they focus on in-person services, which tend to be unpredictable (since people are unpredictable).

A third class of shielded occupations is in education. Like healthcare providers, education workers tend to be involved in providing customized services to people in settings where eye-to-eye contact is important to the service's effectiveness. These professionals include all manner of teachers—primary, secondary, special education, and postsecondary teachers and instructors.

The arts, entertainment, and leisure industries also have a lot of shielded work to offer, since personal contact is so often an essential aspect of the service provided. This includes occupations like craft artists; floral, interior, and exhibit designers; and coaches and scouts. It also encompasses performing artists like dancers, choreographers, actors, musicians, and singers.

As mentioned, this list of jobs should be viewed as drawing a line-sketch portrait of the jobs of the future. Most of us will work in jobs that resemble but are not actually these jobs. In 1850, for example, the future of work was clear in its general outlines, but not in its details. Sixty percent of people worked on farms in the US and it was clear that this share would fall drastically. It was also clear that the new jobs would be in manufacturing and services, but it was not at all clear exactly what the new occupations would be.

While we don't know the names of the millions of future jobs that will be created to replace those taken by AI and RI, we can think about the sort of economy that the new jobs will create.

TOWARD A MORE LOCAL, MORE HUMAN, COMMUNITY-BASED ECONOMY

Sherlock Holmes, the fictional Victorian sleuth, said: "When you have eliminated the impossible, whatever remains, however improbable, must be the truth." This is the principle we should use when thinking about what our lives will be like after the Globotics Transformation. Future jobs will rely heavily on skills that globots don't have.

Direct wage competition is not a feasible way to combat job displacement. White-collar robots are happy with zero wages, and many foreign remote workers will work for very little. We cannot plan on keeping the jobs that globots can do. The jobs that will be left—and the masses of new jobs that will be created by boundless human ingenuity—will be in areas that are sheltered from globots. This will transform lives. It will reshape economies and communities.

When people moved from farms to factories, and then from factories to offices, communities changed. The same will happen again. My guess is that it will make for a better society. My guess is founded on three clues. First, the jobs that will be left will be those that require face-to-face interactions. This will make our communities more local, and probably more urban. If you really do have to go into the office every day, there are big benefits to living near your place of employment.

Second, the jobs that thrive in the face of AI competition will be those that stress humanity's great advantages. Machines have not been very successful at acquiring social intelligence, emotional intelligence, creativity, innovativeness, or the ability to deal with unknown situations, so the human jobs of the future will involve doing things for which humanity is an edge.

Third, once we manage the transition to new jobs and new sectors, the globots will make us richer. Things made cheaply by globots will cost less for humans and this will make us materially better off. The globotics revolution could mean soaring productivity that could finance a breakthrough to a new nirvana, a better society that offered fulfilling work and fostered more caring-and-sharing attitudes. Think of *Downton Abbey* where all the servants are globots. Adding breakthroughs in medicine and bioengineering into the mix means that our lives could be very long as well.

Combining these three streams of guesses about the future suggests another stream of guesses. The result could be a new localism—a trend that should reinforce local, social, family, and community ties. Understanding this leap of logic requires a quick dip into social anthropology—the field that studies why different societies are so different.

The departure point is the so-called social dilemma. Individuals tend to be individualistic, but achieving outcomes that are good for all of us usually demands that we dial down our selfishness. Joshua Green, a professor of psychology at Harvard, refers to this dichotomy as "the fundamental problem of human existence."[11] Our success and happiness require a pursuit of collective interests, but evolution tends to reward self-minded individuals who free ride on the community. The prime directives of societies are designed to solve the fundamental problem. Successful societies are those whose social fabric and institutional organization "square the circle" when it comes to this me-versus-us issue.

Green maintains there are two basic forms of "kinship systems" which provide two very different solutions to the fundamental problem. One set of societies solves the problem with strong group-ish-ness. In the extreme, this means highly organized, cohesive groups that have dense social networks. Think of village-like communities where everyone knows

11. Joshua Greene, *Moral Tribes: Emotion, Reason and the Gap Between Us and Them* (London: Atlantic Books, 2014).

everyone and all of their relatives. This is the "kith and kin" solution. Another solves the problem with external constraints that coordinate and redirect individualism. These include the shaming of antisocial behavior based on religion, morality, or formal laws.[12] Most societies rely on a blend of the kith-and-kin and external-constraints solutions.

A more local, more human society that seems to be on the other side of the globotics upheaval is one where the kith-and-kin solution rises in prominence compared to the external-constraints solution. The point is that frequent, in-person exchanges help create kinship bonds. Another guess in this line of guesses is that the extra wealth will make it easier for us to all get along. A society where material well-being is widespread is a society that has smoothed off many of the hard edges of the me-versus-us dilemma.

Straight-lining this thought into the future suggests that our more local, more human workplaces will foster more cohesive and supportive communities. The last guess in the string of guesses is about locality preferences. The tendency to buy local could rise. The new material affluence and the new localism of communities could create what might be called the "handicrafts economy." We already see a preference for made-local things—at least among the people who can afford them. Handmade beer, to pick a product for which localism is rampant in the US, is reflective of the trend. People pay more for local craft beer more or less exactly because it is made in such an "inefficient" manner. Small batches brewed without automation, using expensive ingredients, and drawing on human creativity result in pricey, but oddly attractive adult beverages.

These points, taken together, are why I am optimistic about the long run, why I believe the future economy will be more local and more human. The sheltered sectors of the future will be where people actually have to be together doing things for which humanity is an edge, not a handicap. This will mean that our work lives will be filled with far more caring, sharing, understanding, creating, empathizing, innovating, and managing—all

12. For evidence on this, see Benjamin Enke, "Kinship Systems, Cooperation and the Evolution of Culture," NBER Working Paper No. 23499, 2017.

with people who are actually in the room. The sense of belonging to a community will rise and people will support each other.

All this is wild speculation, of course, but I don't think it is wild to suggest that the Globotics Transformation will eventually alter our way of life as fundamentally as the Great Transformation altered lives in the nineteenth and twentieth centuries.

How should we prepare ourselves and our children for the positions that seem likely to thrive in the Globotics Transformation?

10

The Future Doesn't Take Appointments: Preparing for the New Jobs

At a June 2017 promotional event in New York, Amelia came face-to-face with Lauren Hayes—the human model on whom Amelia's avatar is based. Or actually, it was face-to-screen since Amelia is a piece of software that only lives inside computer equipment.

In a rather heart-warming stunt, Amelia's maker, Chetan Dube, staged a quiz show between Hayes and Amelia. The human won. Hayes easily responded to general quiz questions faster than Amelia and with more natural language. Of course, the contest would have gone very differently if the questions had been in Swedish and the topics had focused on opening bank accounts.

This quiz-show could be taken as a metaphor for the entire Globotics Transformation. Companies will be running contests between humans and globots in the years ahead. Sometimes the humans will win; sometimes the globots will win. In this case, Hayes's win was based on one of humanity's greatest advantages—general intelligence and an ability to deal with new situations.

There are important clues here as to how we should prepare for the age of globotics.

The Old Rules Are Aimed at the Old Problem

Every economic transformation creates triumphs for those who can seize the opportunities and tragedies for those who can't. Preparation is essential. One very obvious way forward is to return to the analysis of the capabilities of artificial intelligence (AI) and remote intelligence (RI) while keeping in mind the advantages of having real humans in the same room. In a nutshell, preparation should focus on enhancing people's strengths in areas where neither AI nor RI are strong, and avoiding large investments in skills where AI or RI will soon rain down a fury of competition.

This brings us to the first fundamental rule for thriving in the age of globotics: the old rules won't work.

The most prominent of the old rules was a simple dictum: "Get more skills, education, training, and experience." This formed the backbone of many national strategies and the thinking of many families worried about their children's future prospects.

The old rule did make sense before digitech. It rested on the bedrock fact that the disruptive impacts of automation and globalization were limited to sectors that involved making things—manufacturing, agriculture, and mining. Services, by contrast, were naturally sheltered from automation and globalization since computers couldn't think, and most services were very hard to trade across international borders.

Given this, the old rule worked for a very simple reason. Having higher skills and higher education made it more likely that you'd get a job in a sheltered service sector rather than a goods-producing sector that was exposed to automation and globalization. The old rule helped people avoid competition from industrial robots at home and China abroad. And it helped them seize the opportunities created by Information and Communication Technology (ICT) in the service sector.

Getting more skills made it more likely that you'd get a job on the winning side of the "skill twist." ICT produced a type of automation that acted as a better substitute for people who worked with their hands, while making better tools for people who worked with their heads. The old rule was the best way of getting on a glide path that took you to a job where ICT was a helper, not a hurter.

Until the digitech revolution took off, especially machine learning, most service-sector and professional jobs were shielded from automation since industrial robots could not speak, listen, read, write, or help around the office in any way. Likewise, competition from foreign service workers was an issue for, say, back-office tasks like processing expenses or updating customer accounts, but the range of offshorable office jobs turned out to be rather restricted given the limits of telecommunications and the difficulty of coordinating with remote teams. In short, higher education was the ticket to getting out of the goods-making sectors and into the service sector. This won't work any longer.

The digitech revolution repealed the old reality on which the old rule was based. Many formerly sheltered jobs in the service sector are now "ground zero" for the Globotics Transformation. And this means that the "get more skills" advice is too blunt for today's world. Simply getting more skills and higher university degrees will not take you out of the job-wrecking path of AI and RI. The disruptive aspects of the globot revolution are focused firmly on previously sheltered service jobs. The eruptive pace of digital technology is making white-collar robots very good at helping around the office, and very capable of taking over many of the tasks that are now done by people who work with their heads.

Digitech is also rapidly making it easier to slot remote workers into local teams. The main thrust of this so far has been to allow domestic workers to work remotely. But increasingly, the same changes will allow foreign remote workers to be slotted into local teams. The inevitable result is that domestic workers will face new competition from talented foreigners sitting abroad and willing to contribute their skills for little money. It will bring many service-sector workers in the advanced economies into direct wage competition from workers in emerging economies.

This is why the old rules will no longer work. Globots are threatening jobs in the service sector where three-quarters of our citizens make their living. Preparing for the Globotics Transformation will require a different way of thinking.

THREE RULES FOR THRIVING IN THE AGE OF GLOBOTS

Nothing has changed when it comes to radical changes—they create more opportunity for some and more competition for others. It's all down to preparation. Three rules will help prepare ourselves and our children for the globotics revolution. These are just common sense. First, seek jobs that don't compete directly with white-collar robots (AI) or telemigrants (RI). Second, seek to build up skills that allow you to avoid direct competition with RI and with AI. Third, realize that humanity is an edge not a handicap. In the future, having a good heart may be as important to economic success as having a good head was in the twentieth century, and a strong hand was in the nineteenth century.

The first rule tells us to move away from skills that draw solely on experience-based pattern recognition, since AI is getting very good at such things. Machine learning has pushed the capacity of computer automation far into cogitative territory that was previously a no-go zone for computers and white-collar robots. If it is possible to gather a big data set on a particular task, that task will soon be taken over by AI-trained software robots. Try to stay away from jobs where that has, is, or soon will happen.

Likewise, we should move toward skills that help us deal with real people who have to be in frequent in-person contact, since that is something telemigrants can't do. Digital technology—especially advanced communication technologies, machine translation, and online international freelancing platforms—are making is easy for talented, low-cost foreigners sitting abroad to undertake many tasks in our offices. Which tasks are these? One obvious set of clues lies in the tasks that are today done by domestic workers telecommuting part-time or full-time. Try to stay away from jobs and tasks where you don't actually have to be in the room with others; these are the tasks and jobs where you will soon be competing with educated foreigners who can support a middle-class lifestyle on $10 an hour.

In terms of training, we should invest in building soft skills like being able to work in groups and being creative, socially aware, empathic, and ethical. These will be the workplace skills in demand because globots aren't good at these things.

Of course, it can't be 100 percent soft skills. We will all have to be more technically fluent—but that is already true of most people under thirty today. One point that is often lacking in the public debate is as simple as it is obvious. Most people who win from the Globotics Transformation will be *using* globots, not *designing* them. A few AI and telecommunication experts will get fabulously wealthy, but that is an irrelevance in the world of work. Putting it starkly, if you don't want to be replaced by globots, you will probably have to learn how to use them as tools in your job.

Flexibility and adaptability will surely be important in the fast-moving, future world-of-work. Language skills, by contrast, will provide less of an advantage than they did before machine translation got so good.

Consider an example of how globots changed the meaning of success in the law profession. Until recently, a law degree and a can-do attitude was a ticket to middle-class prosperity. Now, junior lawyers are competing with white-collar robots; those who can leverage the new tech may thrive, but those who can't will have to find something else to do.

The Legal Jobs Example

Berwin Leighton Paisner is a British law firm that works on property disputes. In the past, they threw junior lawyers and paralegals into a room with hundreds of pages of documents from which they were expected to extract critical data. That created weeks of work for young, on-their-way-up lawyers. Now, the firm uses an AI system that extracts the same information in minutes.

Christina Blacklaws, director of innovation at another UK law firm and president of the Law Society of England and Wales, notes that law students need tech skills, not just law skills: "Most universities continue to teach a traditional curriculum, which was fine up until a few years ago, but might not properly prepare young people," she notes. Law students will have to train themselves.

There are also hints of rule number three (humanity is an edge, not a handicap) in Blacklaws's advice. Robo-lawyers don't run themselves. They are to tomorrow's lawyers what a plow is to farmers today—a handy tool that magnifies your usefulness if you know how to use it. Human lawyers can do many things robo-lawyers can't; turning this insight into income, however, requires investing in particular forms of knowledge.

Another case study in the three rules comes by looking at the way modern corporations are creating the future of work.

The Agile Teams Example

Something deep is going on in modern companies—digital disruption is what many call it. With technologies and competition accelerating, service-sector companies are shifting to more flexible organizational models. That means more flexible arrangements with workers. They are blending in-person jobs with RI and AI in ways that allow employees to be "agile" and use this advantage to disrupt traditional corporations that continue to employ on-the-spot workers to do most things.

In the not-too-distant future, AI and RI will allow smart, dedicated, in-place, and flexible teams of generalists sitting in the same building to direct much larger teams of telemigrants and white-collar robots. This combination of in-person, remote, and synthetic workers will allow the teams to react quickly to new opportunities and quickly retreat from failures. One buzzword for this is "agile."

"Agile methodologies—which involve new values, principles, practices, and benefits and are a radical alternative to command-and-control-style management—are spreading across a broad range of industries," according to management specialists Darrell Rigby, Jeff Sutherland, and Hirotaka Takeuchi.[1] When a new challenge arises, companies using the agile-team approach creates a team of from three to nine people who have the necessary range of skills to seize the opportunity. Agile teams manage themselves but are fully accountable for what they do. The biggest winners from the

1. Darrell Rigby, Jeff Sutherland, and Hirotaka Takeuchi, "Embracing Agile," *Harvard Business Review*, May 2016.

Globotics Transformation will be the members of these smart, dedicated, in-place teams. For them, globots will act as new tools, not new competition.

These conjectures are about how people can prepare. A separate question is: What can governments do to help?

PREPARING FOR THE UPHEAVAL — PROTECT WORKERS, NOT JOBS

Change is difficult, especially when it comes fast and seems unfair. If the globotics upheaval leads to violence or radical reactions, it will be because of the trend's velocity and injustice. To make such outcomes less likely, governments need to help workers adjust to the job displacement, foster job replacement, and—if the pace turns out to be too great—slow it all down with regulation, and Employment Protection Legislation.

The iron law of globalization and automation is that progress means change, and change means pain. As Pascal Lamy, a man who spent years dealing with the backlash against globalization in his role as director-general of the WTO, puts it: "Trade works because it is painful, and it is painful because it works."[2] The exact same thing applies to globotics. An extra dollop of political difficulty is added by the fact that globalization and automation often favor those who are already favored.

The best way to address this conundrum is to reinforce policies that make it easier for people to adjust. Governments who want to avoid explosive backlashes must figure out how to maintain political support for the changes. They will have to find ways of sharing the gains and pains.

While redistributive policies will undoubtedly be part of the solution, they can only be a temporary fix given how people's lives and membership in communities are defined by they jobs. The

---

2. Pascal Lamy, "Looking Ahead: The New World of Trade," speech at ECIPE conference, Brussels, *ECIPE.com*, March 9, 2015.

flexicurity policies in Denmark are a good inspiration for what is possible.[3]

Danish flexicurity rests on a triangle of policies. The first is a policy of allowing firms to easily fire and easily hire workers. The second corner is a comprehensive safety net for workers who lose their job. Unemployment benefits are generous but only at moderate income levels; they replace about 90 percent of the wage, but only up to a maximum of about $2,000 per month. The last corner is "activation" policies, which means things that help displaced workers get new jobs. These policies range from job-search assistance and counseling all the way to retraining and wage subsidies.

Much more could be said about government policy, but in my view nothing novel is needed. Economic transformations have been forcing people to change jobs since the industrial revolution. Different governments have tried different policy mixes to help their citizens adjust to these transformations. Some nations have been successful at this—those in Northern Europe and Japan are good examples—but others have not.

I cannot see how the Globotics Transformation adds anything new to the solutions needed—except that it will all come much faster, so the need for Danish-style labour-market adjustment policies will be even greater in the future than it was in the past.

My guess is that the nations which were most successful in navigating the upheaval experienced since 1973 will be the same ones that succeed in avoiding extreme backlashes during the globotics upheaval. I am particularly worried that America's reliance on rugged individualism will produce outcomes that are especially rich for rich citizens, but especially rugged for average citizens.

3. For more detail see Torben Andersen, Nicole Bosch, Anja Deelen, and Rob Euwals, "The Danish Flexicurity Model in the Great Recession," *VoxEU.org*, April 8, 2011.

CONCLUDING REMARKS

Technology and more internationally open markets can produce outcomes that are good or ghastly. It is mostly a matter of speed. The past provides important clues on how we can make the outcomes good and avoid having them get ghastly, so a quick recap of the historical experience is useful.

The tech impulse behind the Great Transformation was steam power. Steam took the horse out of horsepower and put horsepower into man-power. It was like giving people massive muscles. It allowed humans to control and concentrate previously unimaginable amounts of power. Mostly, this created better tools for people who worked with their hands. A century later, steam launched modern globalization.

The impulse launched the economy on a very rocky, three-century ride that covered two world wars, the Great Depression, and the rise of fascism and communism. After populist leaders like FDR in the US and Clement Attlee in the UK introduced "New Deal" social welfare programs, the Great Transformation started to be a great thing for the majority. Income inequality fell.

A very different tech impulse started the Services Transformation from 1973 or so. Miniaturization of computers fired the starting gun on a slew of innovations that made it cheaper and easier to process and transmit information.

This ICT revolution had two very different effects on the world of work. First, it took the "man" out of manufacturing by allowing robot "hands" to do things that previously only human hands could. Second, it put powerful tools into the hands of people who worked with their heads, thus massively multiplying their mental "muscle." It allowed office workers to control and process previously unimaginable amounts of information. Two decades later, ICT launched the "New Globalization" where firms took their know-how abroad and combined it with low-cost labour in a way that further undermined the fortunes of factory workers.

The ICT impulse launched the economy on a very uneven ride. The resulting deindustrialization and shift to service jobs were devastating for some and delightful for others. People who worked with their hands found that the technology devalued their value added; people who worked with their heads found the opposite. Income inequality rose.

A general sense of vulnerability and uncertainty spread since this tech-trade team affected the economy in a very different way it did before 1973. The changes hit the economy and employment patterns with a finer degree of resolution; it wasn't sectors and skill groups any more. The changes happened at the level of production stages and even individual jobs.

The Globotics Transformation was launched by digital technology that differs from ICT in subtle yet important ways. Oversimplifying to make the point, ICT replaced those who worked with their hands and rewarded those who worked with their heads. Continuing to oversimplify, digitech is replacing people who work with their heads and rewarding those who work with their hearts.

Tasks that involve routine manipulation of information will be taken over by globots. Globots won't take over tasks where humanity is an edge or tasks where being in the same room is essential; these tasks will be sheltered from automation and globalization in the future world of work.

The resulting shift into sheltered service and professional jobs will reward a very different set of skills than the skillset that ICT rewarded. Ultimately, artificial intelligence will make everyone a lot smarter in the IQ, pattern-recognition sense of the word "smart." The change will be revolutionary for average people, but much less so for the few who are very clever to begin with.

Using "head" in the sense of "brain", AI will give more "head" to people with big hearts, but no extra heart to people with big heads. I think this twenty-first century skill twist will have unexpected implications for income inequality going forward. Presuming that the distribution of "heart" skills in the population is basically unrelated to the distribution of "head" skills, there is no reason that this new skill twist should lead to further rises in income inequality. It might even lower inequality in the long run.

Reaching this felicitous future is the challenge. There is a very real danger that the shift from unsheltered service jobs to sheltered service jobs happens too fast. The danger is that communities feel overwhelmed and push back in destructive ways. If the anger of the displaced blue-collar workers fuses with the anger of the soon-to-be-displaced white-collar workers, the outcome could be backlashes of the 1930s type.

But there is nothing inevitable about this.

It's Our Choice

Computers, air travel, and the postwar opening of world trade transformed societies, but the changes were spread over decades. Each change agitated communities and whole societies by creating new opportunities for some and new competition for others. Each brought with it strong social and economic tensions since—by and large—the new opportunities spurred the fortunes of nations' most competitive workers and firms, while the extra competition harmed the fortunes of nations' least competitive firms and workers.

In recent decades, societies and communities have had time to adjust, so while we have seen abundant disruption and pain, we have not seen radical backlashes. We saw Brits vote for Brexit, and America elect Donald Trump, but truly radical figures have not gained prominence. We have not witnessed the rise of twenty-first century versions of Mussolini, Hitler, or Stalin on the dismal side, or FDR and Attlee on the hopeful side. But it hasn't always worked out this way.

The radical transformations that came with the industrial revolution and the shift from feudalism to capitalism destroyed the social fabric that had, for centuries, been based on reciprocity and ancient hierarchical relationships. As Karl Polanyi wrote in his 1942 book, *The Great Transformation*, the commoditization of labor and mass migration to urban and industrial areas disturbed traditional values to such an extent that the people pushed back by embracing communism or fascism. Back then, however, the push and pushback both took many decades. The industrial and societal revolutions started accelerating around 1820, but communism and fascism took off only in the 1920s.

Things are moving much faster this time. My guess is that it will all work out well in the long run, but only if we can make sure globotics advances at a human pace, and the disruption can be seen by many as a decent development.

This is why it is critical to realize that the pace of progress is not set by some abstract law of nature. We can control the speed of disruption; we have the tools. It's our choice.

INDEX

Page numbers followed by *f* and *t* refer to figures and tables, respectively.